HOW TO SAVE MONEY BUILDING YOUR OWN HOME

HOW TO SAVE MONEY BUILDING YOUR OWN HOME

Selecting the Site · Planning the Dwelling
Estimating the Costs · Negotiating Loans
Acquiring Permits · Complying with Building Codes
Excavating the Foundation · Purchasing Materials
Obtaining Discounts · Framing & Roofing
Installing Plumbing, Heating & Electricity
Supervising the Skilled Tradesman
Finishing the Job

S. Helen Brown & James R. Brown

DRAKE PUBLISHERS INC. NEW YORK · LONDON

ABOUT THE AUTHORS

Helen, Jim, and their son, Alex Brown, live in Sugarcreek Township, Ohio, in the fourth house they have built as their own contractors. Jim is project superintendent for a commercial builder in Dayton, and Helen is an accountant.

Published in 1978 by
Drake Publishers, Inc.
801 Second Avenue
New York, N.Y. 10017

Library of Congress Cataloging in Publication Data

Brown, S. Helen
How to save money building your own home.

Includes index.
1. House construction—Amateurs' manuals.
I. Brown, James R., joint author. II. Title.
TH4815.B75 696 77-18412
ISBN 0-8473-1692-0

Design: Harold Franklin

Printed in the United States of America

CONTENTS

HOW TO SAVE MONEY BUILDING YOUR OWN HOME

Preface

About This House

This comfortable ranch home near Dayton, Ohio was built by the owners, acting as their own contractors. Constructed on a half-acre lot, it is airy and spacious and would fit into almost any neighborhood in the country. Most important, it sold quickly when the owner-builders decided to move, and counting appreciation over two years, netted them a profit of nearly $20,000.

We built this house in 1972, on a lot a little bit east of Dayton, Ohio. The lot's cost was $4,500 which we paid in cash, and we secured a $16,500 construction loan from Citizens Federal Savings and Loan, using the lot as collateral.

The loan was set up on a draw basis, meaning the lender paid part of it each time as needed during various stages in the house's construction. We were required to put up $2,000 as a down payment on the construction loan and had additional expenses of about $1,000, making the total cash cost of this home about $24,000.

We began the physical construction of the house on April 7, 1972 and moved into the finished dwelling on Labor Day weekend that fall. In February, 1977 we sold it for $44,000, which gave us a gross profit of $20,000.

And we moved to another new home a few miles away which, of course, we had also built for ourselves, using the same advice we present in this book. It's fun, it's profitable, and almost anyone can do it if he is blessed with determination, patience, and a spirit of adventure.

Helen and Jim Brown

Chapter 1

Introduction:

Making the Decision to Build

Your old home is just too small. Those two bedrooms have begun to look more like closets, and there is no family room at all. Or the apartment had no place where your children could play. Or the city has passed a new pet ordinance and you need a fenced-in yard. In any event, you need a new home.

But alas, the price is out of sight! That $25,000 home you looked at five years ago now costs $40,000. Homes in some areas have appreciated as much as 20 percent a year. The cost of materials and labor is the reason.

Maybe you have put off buying a home because of higher interest rates. It is very unlikely that labor, material costs, or interest rates will go down in the near future; to the contrary, indications are that they will go up. Now is the time you should be building that home.

If you have looked at the possibility of

Rear view of the Dayton house, a simple ranch design.

purchasing a new home, you may find there is a waiting period of as much as one year before your new home is completed. But you remember that five years ago you paid your down payment, and at the end of three months you moved into your new home. At present there is such a demand for housing in some cities that a ninety-day completion date is almost unheard of.

The price of a new home may be unbelievable, but the waiting list to buy this home at this inflated price is also unbelievable. So why not build your own home and save from $1,000 to $20,000?

"But I am not a carpenter," is your first reaction. Maybe not, but you can read, speak, and write, and that is all that is necessary to build your own home—plus a lot of determination.

The money you save is like tax-free income for you. In addition you will be able to enjoy the tax preference when you sell your residence, and you can take advantage of capital gains treatment on whatever taxable profit you may realize from the sale. And there are other economic advantages.

The amount of money you save is determined by the amount of work you do for yourself. This is not necessarily strenuous physical work, but how many phone calls or trips you are willing to make to acquire permits, hire your own tradesmen, etc. Are you willing to give up all your spare time during one year to develop your project? If you are, you may be able to save as much as $5,000 on the cost of your new home without ever driving a nail. When considering whether the financial gain outweighs the personal time and effort needed for the project, remember how long it takes you to save $5,000 or $10,000 in the bank. Think about adding $5,000, $10,000, $15,000, or even $20,000 to your net worth in just one year!

How do you start and what do you do? You start by remembering that anything you can do for yourself saves you what you would have to pay someone else to do it for you.

Sure, the contractor is trying to keep his costs down, but he won't make five or six phone calls trying to get the lowest cost of materials in the area. Contractors have standard suppliers, and you can usually beat their prices if you are persistent. The contractor passes his increased cost on to the new homeowner, so he doesn't mind not getting the best price in town. But for the new homeowner, it could amount to a few hundred dollars or as much as a few thousand dollars, depending on all the factors. This may be true of subcontractors as well. The cost of materials is a major factor in the total cost of your new home.

A plan that deals efficiently with the cost of materials and labor is the key factor in building your home at the lowest possible price. Planning your home could take as much as six months and the actual construction only three.

Begin your planning with "efficiency" as a way of thinking when relating to your house project. Keep efficiency in mind in selecting the type of house to build. A ranch-style home over a crawl space is usually the most efficient house to construct. Therefore, put aside all thoughts of building that three-story house with a full basement you have always dreamed of owning. Don't try to build the most unusual house in the country. Keep your design simple and efficient. Simplicity in design means efficiency in construction, and efficiency in construction means savings for you.

The long, sunny porch
doubles as a front walk on the Dayton house.
The decorative posts
were installed by the owner, too.

Chapter 2

Selecting Your Building Plans

An open-walled, seaside patio home may be great in Southern California, but the majority of the people throughout the country prefer a conventional home. If you elect to build an unusual home such as an A-frame, be prepared for more difficulty reselling your home. Unconventional homes will also stay on the market longer before someone buys them. Again, keep your home design simple and efficient.

Single story homes with straight lines are the easiest to construct. These ranch-style designs are also the most efficient to construct.

Long, range-type homes have several different exterior styles which are very attractive. Changing from a gable end to a hip roof produces a styling change with many possibilities. The same basic home can be either French or Spanish colonial or modern. The outside of your home can be any style that you desire.

Another important factor in selecting a house plan which will be the least expensive to build is to choose one with a minimum of roof lines. Different roof lines going in many directions could create difficult framing problems requiring the services of a carpenter. Material costs are increased; more material used in the roof section increases the overall cost of your home. This can be avoided by choosing a house plan with simple roof lines. Therefore, if you must compromise on some part of your home to reduce the cost of labor and materials, a good place to do it is on the roof.

Ornamental brick treatment like this distinctive corner styling is something you must work out with the bricklayer before he begins work as it is not always included in the house plans. It adds a distinctive touch to the home, but any special work requires extra care, and a fee must be paid for it.

Newspapers, magazines, and construction trade papers offer a good selection of stock home plans. Designed by architecture firms, these plans are then sold to the public through the various publications. In this way you can have a home plan designed by an architect without paying an exorbitant fee for his services.

The "home" section of your local newspaper may feature a different drawing each week, such as the Associated Press "House of the Week." These plans are inexpensive. If you think you like a certain design, order the "baby blueprints" of the house and study them. You may want to order several different designs. In most of the baby blueprint packages you can also receive the materials list for an additional sum of money. This is also a bargain and could give you some advance notice of just what materials you will need for your new home. At this time materials lists are secondary, but you should already be giving some thought to overall cost in determining your design.

If you like the baby blueprints, more detailed plans can be ordered. When you order be sure to allow several weeks for the blueprints to arrive in the mail. Delivery sometimes takes from nine to twelve weeks.

"Women's magazines" also offer building plans that are inexpensive and of good design. Special publication magazines can be purchased on newsstands and have building plans in a variety of designs and costs. These magazines offer helpful information in planning special areas such as kitchens and bathrooms. You can also purchase special magazines on fireplaces, bathrooms, kitchens, and family rooms. The more magazines and booklets on building you read, the more knowledgeable you will be about your project.

After you have selected a home design and ordered your blueprints, drive through areas where new homes are being constructed or have been constructed in the last five years and find your style. If you find a home that is similar to the one you are planning to build, you are in luck. It would definitely be to your advantage to watch all phases of construction. You may also be able to talk to the builders about the work.

If you find a house like yours which has already been completed, stop and ask the owners whether you may look at the house. If you have questions concerning your plans, ask them. This will help to clarify the plans for you. Most homeowners are proud of their homes and are very helpful.

After watching the construction or seeing the completed home, you may decide the plan you thought you wanted is just not workable for you, or you may be doubly convinced that it is your dream home.

Sometimes what looks good to you on paper is not exactly what you want or what you can afford. Maybe the house plan you thought you wanted has too much detailed carpentry work, forcing you to hire a journeyman carpenter and increase the overall cost of your home by thousands of dollars.

In selecting your plans for your new home it is important to consider the width of your home. A workable width is between 24 feet and 32 feet. The wider the main part of your home, the greater the cost. The cost of material increases as the width of your home increases, whether you use 2 x 4s or trusses. Labor cost also increases as the width expands. More labor is needed to construct a wider structure.

A good house design is one where the main structure is no more than 32 feet wide with 24-foot wings on each end running perpendicular to the main structure. This allows you to build the main structure and wings as long as you wish or as long as your building lot will permit.

The style you choose for the outside of your home determines whether it will be all wood or all brick or a little of each. Don't get carried away with an unusual exterior design. This could cost you a lot more money. Also consider the necessary maintenance on the exterior—painting, for example.

If you do want a special item on your brick work, such as arches or coin corners, you should be prepared to pay the extra cost to the bricklayer if you hire one. If you want to lay your own bricks, consider the extra man hours.

Although the structural design of your new home is important in terms of the cost of labor and materials, it is also important to consider the particular needs of your family. Their living pattern will determine the interior design of the house. Building your own home has more advantages than saving money. The plans can be altered to suit the requirements of your family.

When selecting an interior design, consider where the groceries will enter, be stored, be prepared, be eaten, and leave as garbage. When you purchase your groceries and bring them home, will you carry them into your house from the street, your driveway, or garage? The traffic pattern can be made easier if you plan for it in advance. Maybe it will save you several steps to add a door to the design now.

Storing the food once you have brought it into the house is determined by the food itself. Some food needs to be refrigerated while other food must be kept frozen. Foods with a short shelf life need cabinet shelves for easy access. Other foods with longer storage life could be stored in a pantry.

What kinds of food does your family use most? How much shelf space is required for it? This determines the amount of kitchen cabinets you

For a day or two during the installation → of kitchen cabinets, any meticulous kitchen-keeper may worry about whether the place will ever be neat and clean. It will; meanwhile, don't scratch the finish. Sinks, cabinet tops, and finished plumbing come soon after.

need. The design of your home should provide for a freezer, if necessary, as well as a refrigerator. A pantry may also be an important requirement.

When you plan your kitchen, the placement of the appliances becomes part of the original design. Where you prepare your food is important, and the appliances are determined, again, by the requirements of your family. If you have a large family, you may need a stove with six burners. If your family is small, you can install a conventional stove.

The size of your kitchen should also be determined by desires and needs at this time. Once the house is built, it is too late to decide that you need 20 feet of counter space instead of 12.

Now is also the time to design the dining area of your home. Once you prepare the food, how far do you have to walk from the stove to a table or counter where the food will be eaten? Do you prefer a separate breakfast area or dining room for all meals? Decide where they will be located.

After the food has been eaten and there are leftovers, how do you get rid of them? A garbage disposal or trash compactor is probably the best and most efficient way of disposing of garbage. But if you prefer not to use these appliances, be sure that the garbage cans are easily accessible. The placement of sinks and dishwashers should be included in your plans at this time.

Your family's needs are of prime importance in planning the bathrooms. How much privacy do they need? How many bathrooms do you want, and where should they be located? What size bathroom fixtures will you need? These questions should be answered before you plan your home.

Closets, storage areas, hobby centers, and bedrooms should all be considered when you design the interior.

What source of energy will you use for heating and cooling? Heat pump, solar energy, hot water heat, electric wall heaters, natural gas furnace,

Here is the finished kitchen wall with sinks and cabinets installed.

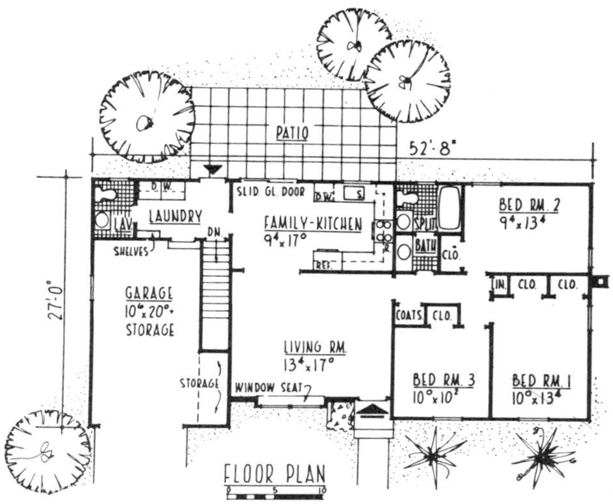

Three bedrooms, two bathrooms, and a patio are some of the features of this ranch-style home designed by York & Schenke, Architects, P.C.

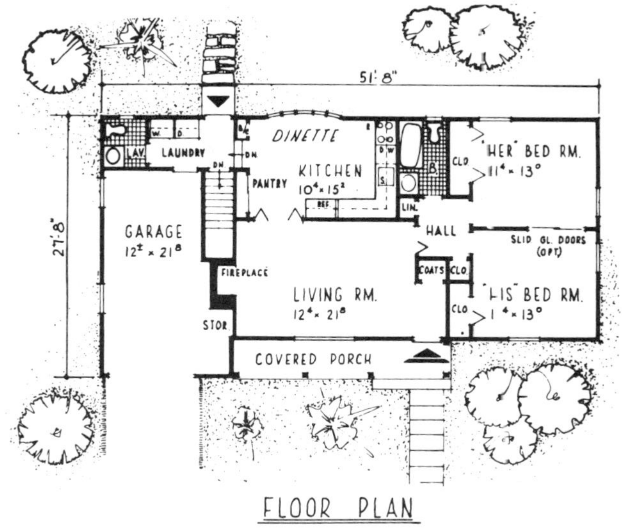

A choice of exterior styles is provided in these plans for a single-story, two-bedroom home by York & Schenke, Architects, P.C.

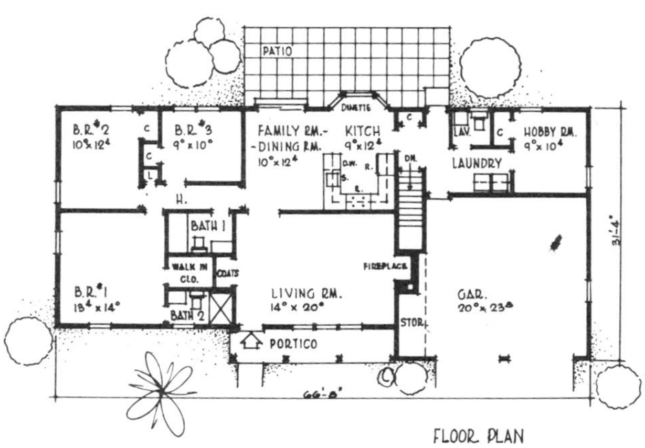

The floor plans for this three-bedroom, ranch-style home by York & Schenke, Architects, P.C., include a fireplace, three bathrooms, and a hobby room.

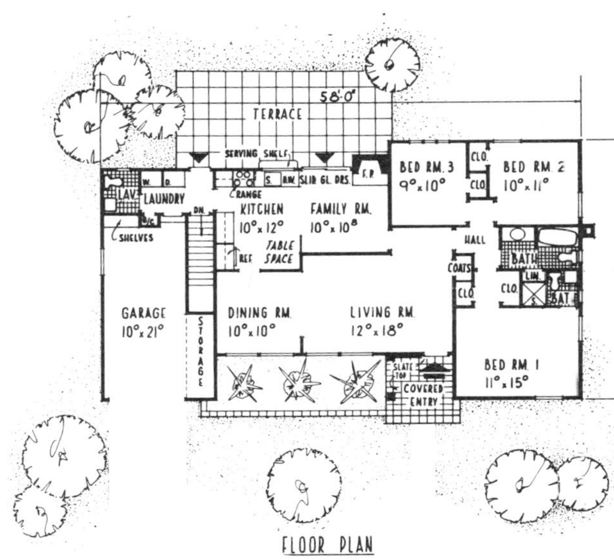

A terrace, garage, fireplace, and family room are attractive features of this three-bedroom ranch home designed by York & Schenke, Architects, P.C.

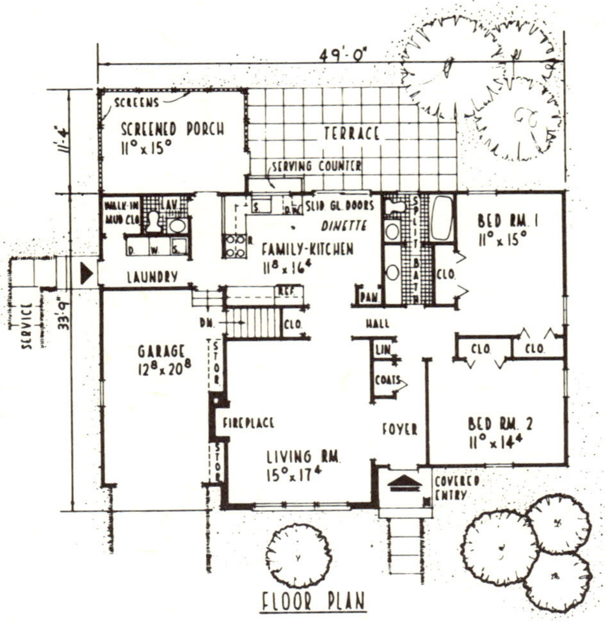

A terrace and screened-in porch increase the ample space of this two-bedroom, ranch-style home by York & Schenke, Architects, P.C.

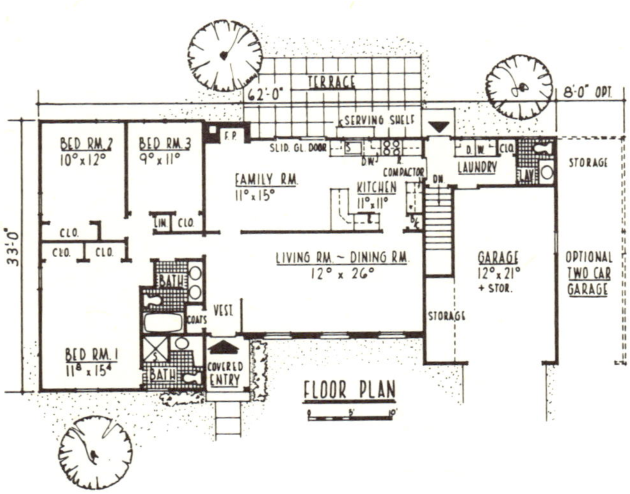

Three bedrooms, a family room, laundry, and optional two-car garage are featured in this ranch home designed by York & Schenke, Architects, P.C.

bottled gas furnace, and oil furnace are some of the choices you have. In most parts of the country, your heating and cooling source is limited to what is available in the area where you plan to build. Many areas have limited the new users of natural gas. Whenever possible, learn all you can about the sources of energy that are currently being used in homes in your area.

As the seriousness of the energy crisis increases, the need to include the most economical heating and cooling system in your building plans is of great importance. A few hours researching the most efficient heating and cooling systems available may save you several hundreds of dollars on fuel and energy cost in the future.

The heating and cooling system you choose will also determine the type of electrical service you need. The service may be 110 volts or 220 volts.

The size of the service also depends on the appliances you plan to use. Most homes now require the 220 volt service. The electrical pattern throughout your home is part of your initial plan. If you want that special outlet for your freezer in the garage, now is the time to put it into your plans. Each light fixture, electrical outlet, and wall switch must be on your plans. If you make a change later, be prepared for additional cost. Don't forget to include your service light fixtures and electrical outlets for front porches and patios.

What size garage do you need? Is a garage something you can do without? Most new homes have provisions for an attached garage. Decide its size and location. Now is the time to include that hobby center or workshop in the garage or an extra light at the garage door. Everything must be included in your plans before you start.

The driveway usually leads into the garage; therefore, when you plan one you should plan the other. Do you need a long driveway, a circular one, or a double driveway?

How many service walks do you need to make your new home more accessible? Think about a future garden or swimming pool that might need walks.

Consider the landscaping necessary to extend your new home to the outside. Your landscaping plans also depend on the building lot you choose.

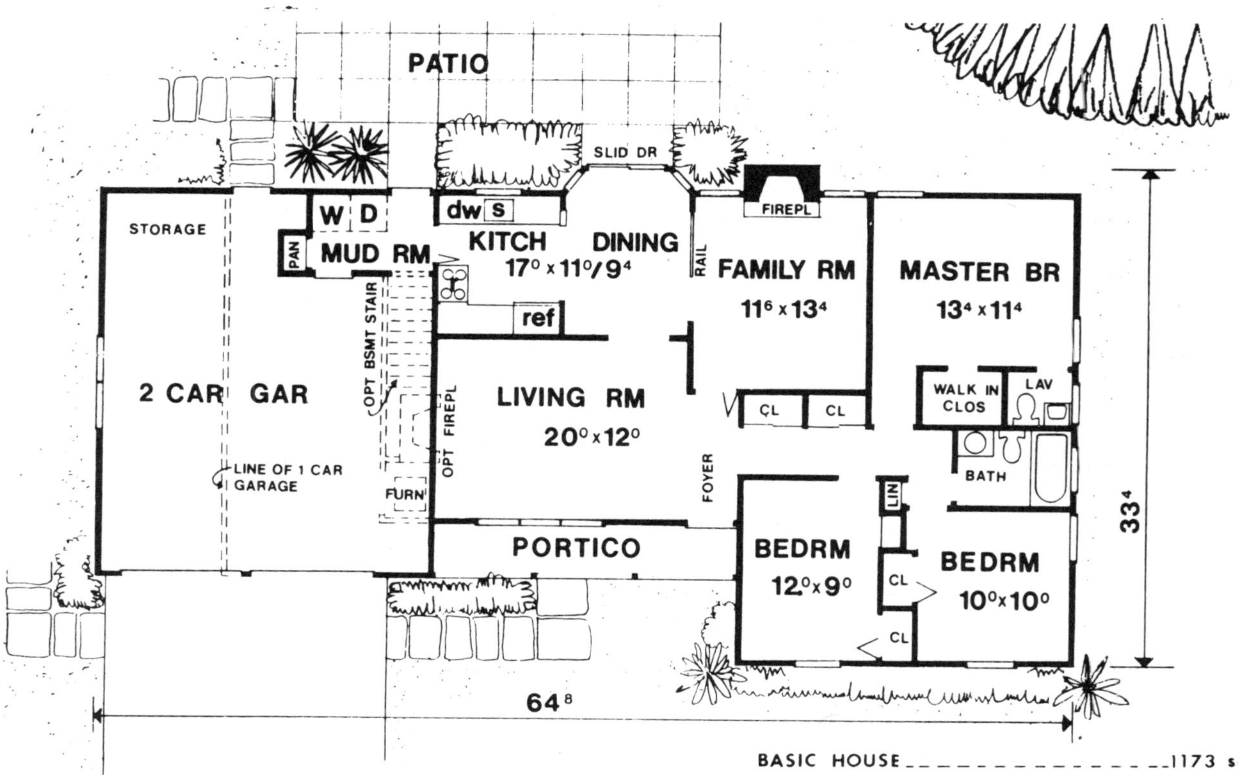

BASIC HOUSE	1173 sq. ft.
MUD RM, UTIL., 1 CAR GAR.	356 sq. ft.
MUD RM, UTIL., 2 CAR GAR.	572 sq. ft.
OPT. BASEMENT	1173 sq. ft.

PONDEROSA RANCH

CONDENSED MATERIAL LIST

For general estimating only. A more detailed list is available with the working drawings.

ONCRETE, MASONRY, STEEL

Footings, Foundation Walls, Slabs, etc.	44 cu. yds.
Optional Brick Chimney	130 cu. ft.
Wire Mesh	1250 sq. ft.

RAMING LUMBER

Total Sills, Studs, Plates, Beams, Rafters, etc.	6858 V.F.M.

HEATHING, INSULATION

3/8" Plywood Wall Sheathing	1670 sq. ft.
3/8" Plywood Roof Sheathing	2185 sq. ft.
3½" Fiberglass Batts	1300 sq. ft.
6" Fiberglass Batts	1210 sq. ft.
1" x 24" Styrofoam	140 lin. ft.

Included with the floor plans for this seven-room, ranch-style home is a materials list for estimating costs. (Blueprints are available from Jerold L. Axelrod, Architect, 275 Broadhollow Rd., Melville, N.Y. 11746.)

WINDOW SCHEDULE

Aluminum Single Hung White 6/6	17
or Aluminum Sliding Bronzed	13

DOOR SCHEDULE

(1) 3-0 x 6-8 x 1 3/4	(1) 1-4 x 6-8 x 1 3/8
(2) 2-8 x 6-8 x 1 3/4	(2) 4-0 x 6-8 x 1 1/8 by pass
(1) 5-0 x 6-8 Alum. sliding	(2) 3-0 x 6-8 x 1 1/8 bi-fold
(4) 2-6 x 6-8 x 1 3/8	(2) 2-6 x 6-8 x 1 1/8 bi-fold
(4) 2-0 x 6-8 x 1 3/8	

EXTERIOR FINISHES

Luan Texture 1-11	350 sq. ft.
Wood shingles	13 squares
Asphalt shingles	22 squares

INTERIOR FINISHES

3/8" Gypsum Board	4500 sq. ft.
½" Gypsum Board	1200 sq. ft.
Ceramic Floor Tile	43 sq. ft.
Ceramic Wall Tile	50 sq. ft.
Vinyl Tile	180 sq. ft.
Carpet	116 sq. yds.

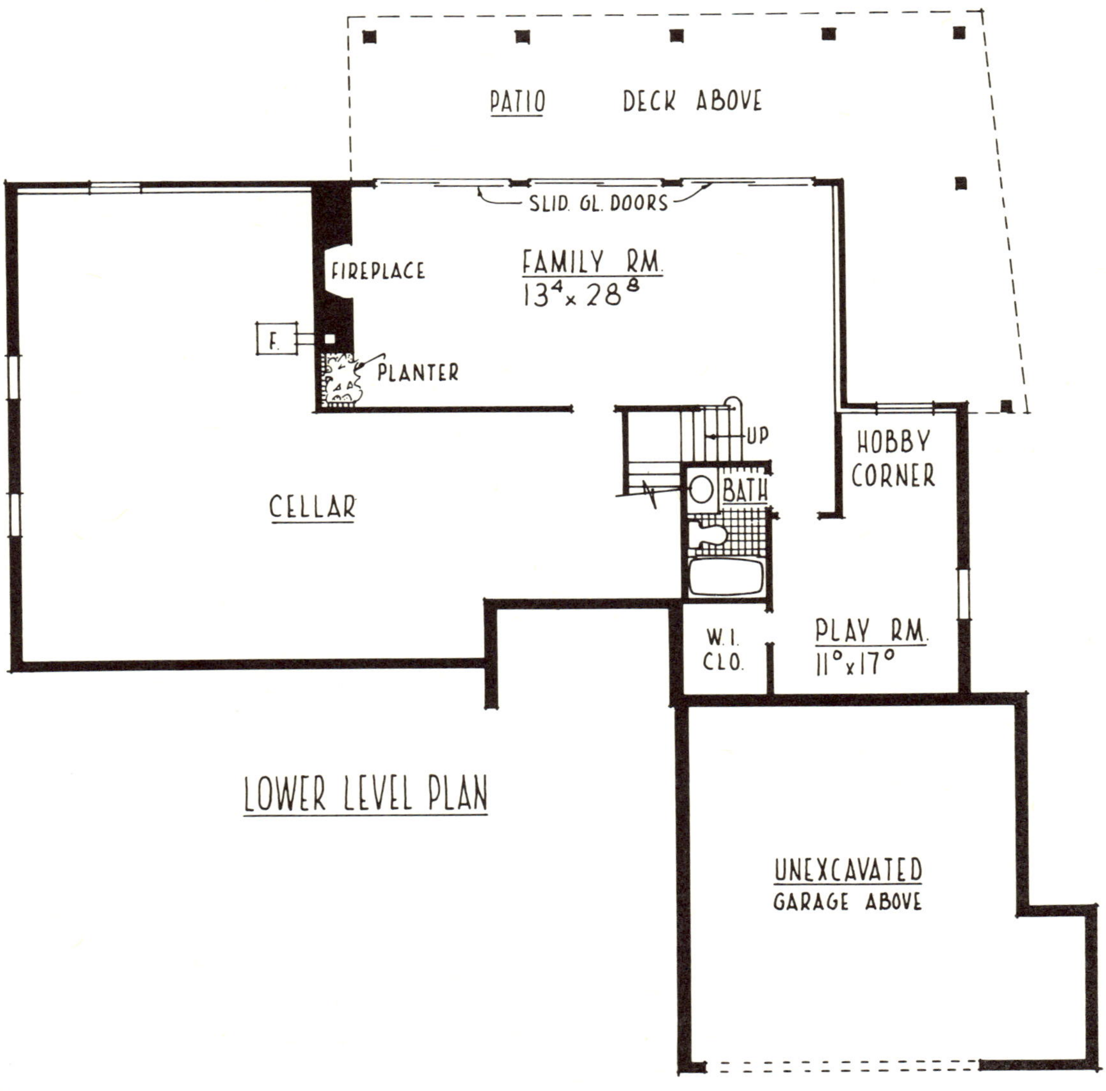
PATIO
DECK ABOVE
SLID. GL. DOORS
FIREPLACE
FAMILY RM.
13⁴ x 28⁸
F.
PLANTER
UP
BATH
HOBBY CORNER
CELLAR
W.I. CLO.
PLAY RM.
11⁰ x 17⁰
LOWER LEVEL PLAN
UNEXCAVATED
GARAGE ABOVE

The floor plans for this two-level house by York & Schenke, Architects, P.C., Garden City, N.Y., include three bedrooms, family room, play room, and dining deck — affording both private and communal areas for the family.

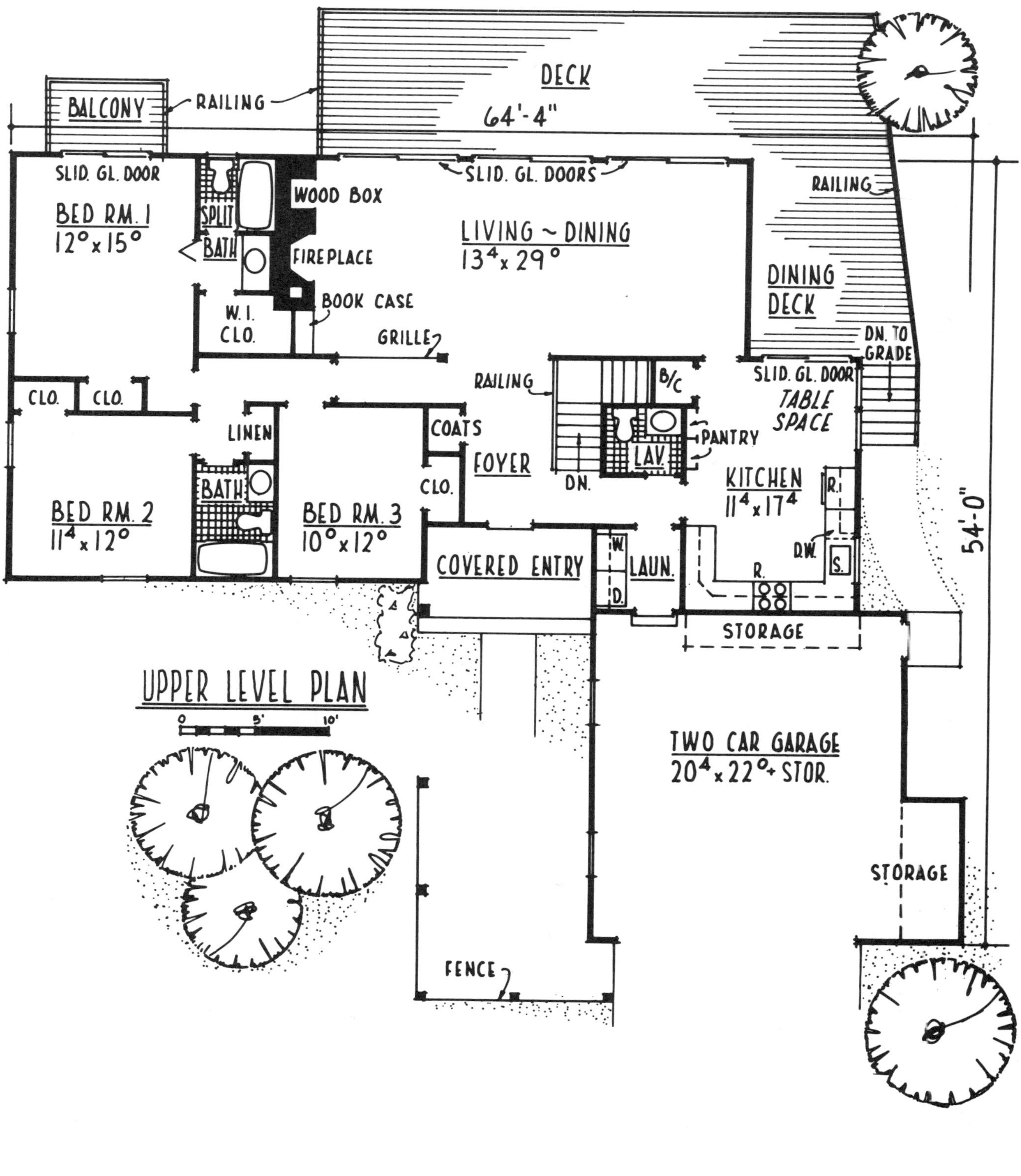

Beautiful lots in nicely shaded settings
in the comfortable community of Centerville, Ohio.
Find what's wrong with this site.

Chapter 3

Selecting Your Building Site

Go out and drive through different sections of the particular area where you think you would like to live. This should be an area in which other new homes are being constructed. The most important thing to keep in mind from the very beginning is: do things for yourself and don't pay someone else to do them for you. Don't pay a real estate agent to find you a building lot. Do it yourself.

There are many advantages to finding your own building lot. While driving around the area you can observe the activities of the neighborhood, locate the schools, make a mental note of playgrounds and parks. You could also make actual notes about the nearest shopping centers, hospitals, highways, fire stations, airports, day care centers, mass transit lines, fast service stores, and other vital services. Locate the nearest freeway and examine the traffic flow at work time.

Once you have found a general area that is of particular interest to you, the next step is to examine the price of the building lots and compare it to other areas. The local newspaper ads will help you determine what is a good price for a building lot in the area. A general rule in buying a building lot is that it is usually 20 percent of the general selling price of homes in the area. Or, on the other hand, it will be about 25 percent of the total cost of your new home.

Most vacant land for sale is advertised in the local newspapers, home advertisers, home finder ads, or simply with "For Sale" signs. Other vacant property may have a real estate sign on the property; in that case you will have to deal through a real estate agent who represents the seller. But don't contract a real estate agent to find property for you. He will only do the same things you could do for yourself. If you find your own building lot, your cost will be minimal, but if you have a real estate agent do it for you, your building lot could cost an additional 6 percent—the agent's commission that the seller added on.

When you have located a particular lot or neighborhood, talk to the residents and ask them about various problems in the area. Why do they like (or not like) the surroundings? Don't hesitate to stop someone mowing his lawn and discuss the reliability of water supply, trash collection, flooding, police and fire protection, or even the existence of little league baseball teams in the area.

Once you have located a desirable building lot and you think the price is reasonable, it is time to contact the tax assessor's office and learn the current tax being charged on that particular lot. While you are in the courthouse you may want to trace the title of the lot. Titles are filed in the county recorder's office and are public records. A title search is usually done by an attorney or a title search company, but if you have the time and are familiar with public recordings of deeds, you are allowed to do it for yourself in most states. An attorney or title company will conduct a title search for $25 to $125, depending on your location and how many times the property has been bought and sold.

Once you have decided that the building lot is desirable and reasonably priced, and know that there are no major tax assessments on the property and you can secure a clear title, ask the owner why the property is for sale and how long it has

This gently sloping, well-drained lot facing a tiny stream looks like a charming site for a new home. Pictures in illustrated real estate bulletins look like this, but . . .

. . . back up a few paces and see the yawning ditch toward which your dream house will face. Its depth indicates a heavy and dangerous flow of water from time to time. Consider this if you have children.

How lovely — a spacious, well-graded lot where sidewalks and street are already in — and in a settled neighborhood. What could be wrong with this location?

. . . This is what's wrong. The rear of the lot and the house that would be built on it face an elementary school. It's convenient if you have children of school age, but you must consider the noise and pedestrian and vehicle traffic. For some homeowners this location would be ideal, but others would disagree.

This lot will make someone a very good setting — on a paved street. Viewing it on a rainy day will give you a clue as to where the low, watery spots are. You can take care of that when the lot is graded.

been on the market. Is this a distress sale due to divorce, illness, or death? Is this building lot suitable for building a residential house? What was the land used for previously, before it was plotted off to be sold for a building lot?

Inquire at your local Health Department about the water and sewer status in the area, particularly with reference to the land you plan to purchase. Will you need to drill a well, and if you do drill a well, how deep will you have to go before you find water? Are you allowed to drill for water on this land, or does someone else own the water rights? If the city, county, or township has a water line that goes by your proposed building lot how much will the tap-in fee be? The local water department or health department will probably be able to answer these questions for you. Also, if there is no central water system in the area, ask your local board of health if there are any plans for a central water system in the future.

You should ask similar questions about the sewer. Ask the board of health if a central sewer is available or if you will need to build a septic tank. If the sewer line does go by the building lot, what will the tap-in cost? If you need to put in a septic tank who will do the testing to determine if the land is suitable for a septic tank? In some counties the local health department's own soil scientist must do this test, but other counties allow outside testing agencies to do it. Ask your local health department the cost of the test and if there is a waiting period before the test can be made. They can also tell you the results of the test in the general area in which you are interested. Ask if there is a ban on building in the area. If there is no central sewer system, ask if one is planned for the area.

Your local health department or regional planning director's office can tell you about the flood plain, and any new highways, hospitals, schools, parks, or shopping centers proposed for your area. You wouldn't want your backyard adjacent to a new shopping center parking lot soon to be developed. Ask any other questions that may concern you about your family's needs.

You should keep in mind that the average quarter-acre subdivision lot is usually not big enough to accommodate a ranch-type home. For maximum flexibility, your lot should be rectangular and at least a half acre.

In most areas there are building codes which restrict the location of your house on your building lot. A common restriction, in many areas, is that you cannot build within 10 feet of your property lines. Therefore, you would need to exclude 20 feet from the building area of your lot. A good size, ranch-style home with a garage could be from 60 to 100 feet long, or longer.

Many half-acre lots will be from 100 to 130 feet wide and very deep. This is due to the expense of street building by the land developer and his desire to get as many building sites as possible on a street or road front. The more home sites he can squeeze together, the cheaper the development cost.

The ideal location of a building lot would be on the north side of the street, facing south, on a street that runs east and west. A house built on a lot like this gives you sun at one end of the house in the morning and the other end in the afternoon with no hot exposure through windows or patio doors. Unfortunately, these lots are often hard to find, so do the best you can in selecting a building lot in the area you want to build your home. If your home must face east or west, look for a lot with shade trees or plan on planting them.

Make a mental note of low spots like this one where someone has obligingly left deep tire tracks. Later you can have it filled.

Know what the land you want to buy was used for in the past. You wouldn't want to buy a building lot that was formerly a dump, junkyard, gravel pit, or used for any other undesirable purpose. Or you wouldn't want to be like the homeowner in Arizona who moved into his new home and discovered an army of small black beetles in the yard. A call to the local exterminator only added to his dilemma, as the exterminator explained the black bugs were cantaloupe beetles, and there wasn't anything reliable to get rid of them. The land had been a melon farm, and the ground was infested.

However, the ingenuity of the wife proved too much for the little black beetles. She put a light out in the yard after dark, and when the beetles came to the light the vacuum cleaner inhaled them. After she collected several vacuum cleaner bags full, she neatly tied all the little black bugs in large green plastic bags, and they were hauled away the next day by the trash collectors—about ten pounds of them!

If the land was used for a landfill, you have to worry about settling. If it was used as a garbage dump, there could be gas leakage from decomposition. If the land was used as a junkyard, the ground could be soaked with grease and oil.

Consider what might be buried there. You'll have enough work making grass grow in the best of soils, but if you're on top of an old gravel pit, you have have to settle for weeds or have to buy tons of expensive topsoil.

It may also be advantageous to know what is under the subsoil, and the best way to learn this is to obtain a geographical survey of the under-

*Natural watercourses do not go away.
If a creek, no matter how small,
crosses your property, then so too will water.
A house on this site
would have to include an underground
aqueduct to move water safely through.
Too much trouble, go somewhere else.*

*In some suburban or rural areas,
lack of ground water means
homeowners rely on cisterns
fed by rainwater
flowing off the roof.
This is sanitary, of course,
if inspected and tested regularly.
Find out about water supply
before you buy property.
Note the splash block at the
bottom of the downspout,
foreground.*

After the county health officer tells you where to place your septic tank, you can buy a prefabricated one ready to drop into the ground.
It has two holes in it, one for the sewer pipe leading from the house and the other one for your drainage lines running out the other way.
If you are joining an existing sewer system, you will not need a septic tank.

ground of your area from your regional planning office.

Your local regional planning director or zoning board can tell you what the land was previously used for and you can determine whether it will be desirable for you.

When purchasing a building lot in the suburbs, be sure to check with the telephone company for local billing procedures in the area you plan to buy. Don't get yourself into a situation similar to that of the couple in Ohio. They purchased a 5-acre tract of land in a township just 3 miles from a very picturesque village, 15 miles from a large city, and 4 miles from the county seat which was in the other direction. Telephone service to the large city, where they were both employed, was a toll call. The telephone number of the local school in the village was also a toll call. Both husband's and wife's work involved getting calls at home, a toll call for most of their clients. Almost all their calls were to or from the city, but the phone company insisted they were in the county seat's telephone area.

When the new homeowners complained to the telephone company about the service, they were told they could have unlimited calls to the city for a fixed fee. But it would cost them nothing to call several small communities as far as 50 miles away from their home in the other direction. The only problem was the couple didn't know anyone in the small communities as far away as 50 miles.

The telephone company explained that the location of their new home was not in a questionable area, but was two miles from the telephone district line change. Although the homeowners' property was located in one town, their telephone service was from the county seat. The homeowners could pay approximately $65 a month, in addition to their regular monthly charge, for a large city telephone number, but they would also have to pay the cost of having the line brought to their property. The homeowners have advised all the people who call them to call collect.

If at anytime you find a building lot that meets every requirement you think it should and you are still unsure of your purchase, call an appraiser for advice. His fee will be minimal compared to the help he may provide.

This black flexible plastic tape is your septic tank drainage line. It has holes on its lower half to allow liquid to flow into the soil.

Just to be sure the drainage holes face in the right direction, the manufacturers of this drain pipe provided a dotted yellow line. The line should face the sky when the pipe is laid. Drainage holes, of course, are on the ground side.

Chapter 4

Purchasing the Building Lot

Now that you have found your building lot in an area that suits the needs of your family at a reasonable price, it is time to purchase the property.

Vacant land is usually considered a bad risk by banks and savings and loan companies and must have some type of structure on it before you can acquire a mortgage for it. The resale of vacant property after foreclosure is very difficult, especially when the resale of the land by a bank or savings and loan is competing against the large land inventory of the developer. Too often, lending institutions have been burned by lending money to unqualified investors who have invested in promotional lake, ski, ocean-front, or desert developments. The vacant land you are interested in buying is for your personal use and is a good investment although the banks and savings and loan associations won't finance it. Therefore, you must have cash to purchase your lot.

There are various ways of acquiring cash to purchase the building lot. One way, of course, is to take it out of your savings account, but not all people have that much money in savings. Another way to get cash is to place a second mortgage on your present home. This extra money is worth the small sacrifice you will have to make in larger house payments. Furthermore, these larger payments will last only as long as it takes to build your new home or sell your old one. You are merely transferring the equity in your present home to your new one.

If you can't mortgage your present home, or don't have one, maybe your credit union would help with your financing. Any type of loan or anything you can do to get money to buy your building lot is worth doing. This is your first step, but make sure you don't use the land as collateral in any way. Your land must be free and clear, and usable as part of your down payment when you apply for your construction loan.

When you have the cash for the building lot, you can then request an attorney to do a title search, or he will refer you to a professional abstractor or title search company who will abstract the property for you. Even if a title search was made earlier, another one is necessary at the time of the purchase. If you also use a title insurance company, they will insure that you are buying the building lot you think you are purchasing. They will also insure the surveyor's plot-plan of the land and that your lot, in particular, has been properly surveyed and recorded. This also protects you against any flaw in the title, against possible errors in the courthouse records of any property, as well as against fraud and other errors that could mean that you don't own the land you have paid for. Title insurance is priced on a sliding scale, depending on the purchase price.

When you buy title insurance for your building lot, you should extend the coverage to the new home you plan to build on the land. This is owner's title insurance. You can also get title insurance on the land now and then later on the house you plan to build. The total premium is paid at one time, and there is no additional cost after that. The insurance terminates when you sell your property.

At the time of sale the deed has to be notarized. That is, the document must be signed by a notary public or public officer who then signs his name, dates the document, and attaches his seal or stamp to the document. The notary public or public of-

16
1 A-BELME
BELMONT-BURGA 1977
11 MORONELL-PERRETTA 1977
4 COZZENS-DUNNING
5 DUNNIINGTON-GATTO 1977
GATTON-HARROW 1977
13 ROSENER-SIPPIO 1977
14 SIPPLE-TABACKMAN 1977
10 MARLU-MORNINGSTAR 1977
121 DAYTON CORP. R 72 REAL 1976
122 DAYTON CORP. R-72 REAL 1977
124 DAYTON CORP. R-72 REAL 1977
127 DAYTON CORP. R-72 REAL 1977
128 DAYTON CORP. R-72 REAL 1977
111
113
114
117

ficer is thereby verifying that the person who signed the document actually signed it and has furnished proof he is the actual person he purports to be.

Why record your deed if you have a clear title to the land, and the title has been properly notarized, and the people are who they say they are? It's because an unrecorded deed is valid only between the two parties, the buyer and the seller, unless it is recorded. The courts only recognize deeds that are of public record, and an unrecorded deed is considered nonexistent.

Recording a deed can save aggravation and expensive litigation later when, for example, the spouse or children of the seller try to take over the property (presumably after his death). Or if the seller loses a lawsuit, your unrecorded property may be seized to pay it off. Or you may find yourself involved in the the seller's divorce case if the public records show he still owns the property. Avoid all this by seeing that your deed is recorded!

Record the deed in the county in which the property is located. Buying property in suburbs often takes you across county lines, leading buyers to think they have to record the property back in the county where they live. Recording your deed in the wrong county is as bad as not recording it at all.

However, there is also a possibility that you may purchase a certain property in one county, and the closing or purchasing will be in another county. In this case you must record the deed in the county where the land is located, and also record your deed in the county in which the closing or the exchange of deed and money took place. Therefore, you may need to record your deed in two different counties. Ask your lawyer.

Your first stop in recording your deed, in many counties, is the tax assessor's office. There you can learn the tax that will be assigned to your property and when billings will be mailed to you. The office personnel will record your deed for tax purposes and probably stamp or sign the deed indicating you have recorded your deed with their office. In some counties it is not necessary to take your deed to the tax assessor's office first for recording, but if it isn't necessary, you should stop at this office just to learn what your taxes will be and the billing dates. Then your next stop is the county recorder's office where you pay a fee to have your deed recorded. This fee varies from state to state, and the cost of recording your deed may depend on the cost of your property or be just a flat recording fee. The name of the recorder's office also varies and could be referred to as the registrar of deeds, title registrar, and/or the county recorder.

In most counties, you pay your fee and receive a receipt. The recorder's office usually records your deed and mails it to you within a week. Be sure to keep your receipt when you pay your recording fee. This is your proof that the courthouse has your deed. Sometimes the people at the courthouse misfile or misplace documents, and if this should happen to you, it is your responsibility to prove they have your deed. It will also help the employees of the courthouse or recorder's office to find your deed or correct any mistakes if an error should occur.

The county treasurer, tax assessor, or court clerk
has public records showing ownership
of property in that jurisdiction.
Files are usually listed by street address, as well as
by the owner's name.
In most areas, this research is free if you come to
the office to do it yourself.
Office personnel will happily help you.

Chapter 5

Estimating Construction Costs

The demand for new housing, strikes by employees of suppliers, natural disasters, and various other factors have made estimating a building's cost a little more difficult. The overall cost of constructing your home could change between the time you start receiving quotes on building materials and the time the home is completed. Nevertheless, when estimating your costs, you must base your computations on current market prices.

You can obtain cost estimate sheets from your local bank or savings and loan company. There are various types of cost estimate sheets available, but remember that you must analyze each individual step yourself before putting them into summary sheets to be shown to the lender.

Your new home costs begin with the plans you have bought. Then the surveying and staking-out of your house on the building lot should be done by someone who knows how. If you know how, and want to do the work yourself, you may still have to rent a transit. List the costs on your estimate sheet.

Insurance is a necessity on your new home from the time you begin planning. Before you do any work on your project, you should consult with your insurance agent concerning the amount of insurance you will need. Construction insurance is usually available and will cover your new home on a step-by-step basis, escalating with the value of your new home until construction has been completed.

You can get a temporary power hookup
from the local electric company
while construction is under way.
Later, the meter will be moved
to the house from the post.
Most installations are now underground.

There are various ways of insuring your new home while under construction. Also ask your insurance agent about on-the-site building material loss and liability coverage. The cost of these policies is minimal, and if you are lucky, you will never need to use them.

In estimating your material it is best to consider one step at a time as if you were actually building the house, and by doing this you are less likely to forget something.

The cost of your building permits may shock you if you are not accustomed to buying them. The best way to get an idea of their cost is to call or visit the building regulatory agent for the area in which you are building, and ask the cost of all permits. Permit costs vary, and the actual buying of the permits will be done later. Right now you are concerned only with finding out their cost.

The next cost to consider is your driveway. Do you need one? Do you already have one? More than likely you will have to build one. Most building lots do not have provisions for a driveway, so to get material from the street to your construction site, you will have to build one. The cost may include removing trees, clearing the site, excavating with heavy equipment, putting down a base for the driveway, and putting in a drain tile to connect you to the main street. In most areas you also have to have a driveway permit so don't forget that. Then estimate whether your cost will include outside labor to finish your driveway or, if you plan to do it yourself, only the concrete or asphalt. You can get prices by calling various asphalt and concrete contractors listed in the telephone directory. You can also find the lowest cost of asphalt material and concrete material simply by calling as many suppliers as you wish until you get a good price. You will find some are rather expensive and others are reasonable. Make several calls to get a comparison.

Did you buy a building lot that requires a great deal of tree removal? Or did you buy a building lot for which you have to purchase fill dirt? This is part of your estimated cost. Get quotes from the excavators, tree service, or other firms you need to hire to clear your building site or fill it in. Again, the best way to contact these people is by telephone. Don't hesitate to ask questions.

You need temporary utility service for your construction site. Call your local power and light company for rates on temporary monthly service. You also need to know the cost of material to build your temporary electric box or the cost of having an electrician install one for you. If you do the work yourself, your only cost is material, and a call to your local electric supplier will give you that information. An electrical contractor will tell you what his charge for installing the temporary service would be. Call several electrical contractors. Never do business with just one quote from one subcontractor. Always try to acquire a minimum of two.

In some areas you may be able to get temporary water service, but it is very rare. Contact your local water department for temporary tap-in fees and monthly rates. In your estimate you would also have to include the cost of a subcontractor to bring the temporary water service to your building site or the cost of the pipe and excavating if you do the work yourself.

Building the foundation footings requires excavating, concrete, and labor. A backhoe operator can be contracted to do your digging, and calls to local excavating contractors will give you his cost. How much concrete you need will be determined by the size of your house and the building requirements in the area. Telephone calls to local concrete suppliers will give you your cost of materials, or you may already have the cost of concrete from wherever you estimated the cost of your driveway. The other cost is labor. If you do the work yourself, forget labor costs. But if you hire someone to do the work for you, his cost must be added to your estimate.

You may prefer to subcontract all the installation of footings to a single subcontractor. In this case you simply ask several contractors to bid, and then select the lowest one.

Cost of foundation walls should be estimated next. Block or concrete are the materials to be used. Is it cheaper to buy concrete blocks and have them laid, or lay them yourself, or build forms and pour concrete? Ask masonry material suppliers in your area the cost of concrete blocks, concrete ready-mix, or mortar mix and sand. These are the materials which will be needed if you use concrete blocks for your foundation walls. Also, who will do the masonry work? If you hire someone to do the work, what will his charge be? Ask. Call several masonry contractors for a quote price. Maybe you can lay the blocks yourself and avoid labor cost.

If you decide to have concrete foundation walls, you need to know the cost and amount of the rough lumber you will need to build forms. Price quotes from your local building supply or lumber company will give you the cost of your lumber. Can you build the forms, or will you hire someone to build them for you? You have the cost of concrete from your driveway estimate so determine which suits you best. You may want to subcontract the foundation walls. In this case, just make your telephone calls to subcontractors and list the best price.

How much will material cost to waterproof your foundation walls? How much labor will be

This is the first page of a four-page form which is very helpful in estimating the cost of a new home.
The next three pages show the balance of the form, which provides columns for listing prices of all materials.
It is almost impossible to obtain a construction loan without a cost estimate.
(Reprinted by permission of the Standard Homes Plan Service, P.O. Box 1919, Raleigh, North Carolina 27602.)

Contractor's Estimate

____
Contractor

____
Address

Office ______________ Home ______________
Telephones

Estimate for: ______________________________

Address: ______________________ City ______________

Phone: ______________ Office ______________ Home

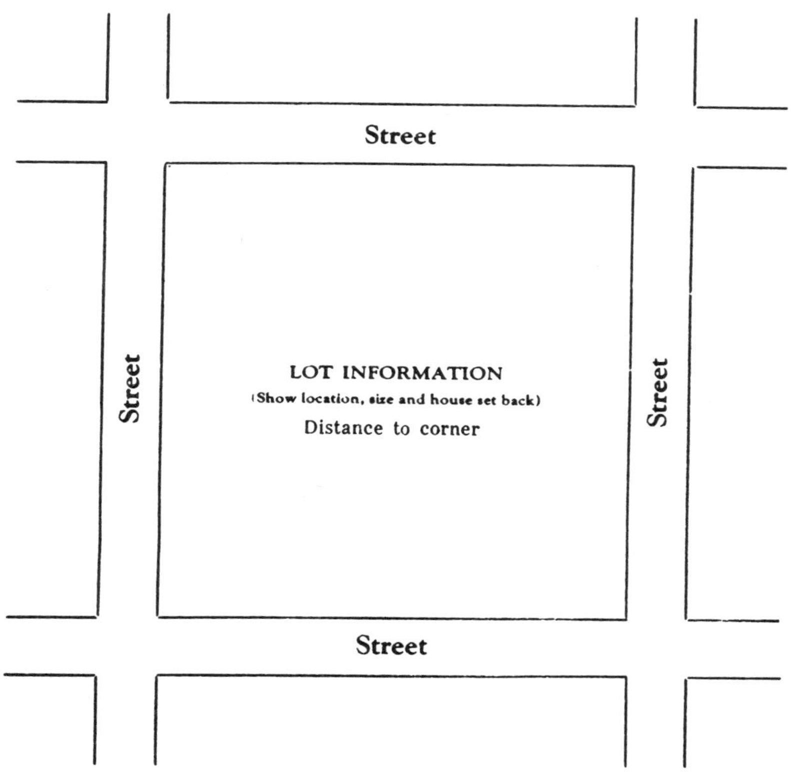

BUILDING SITE

Street ______________

Subdivision ______________

Lot No. ________ Block ________

UTILITIES AVAILABLE

Water ________

Electricity ________

Sewer ________

Gas ________

Paving ________

Front ______________________ Rear

LOT GRADE

This Estimate Based on Current Cost and Supplies and is good not exceeding

a period of days or until............................, 19....

ITEMS	COST ESTIMATES		Estimated Cost	Actual Cost
PRELIMINARY COST	Plans	$	$	$
	Survey and Laying off House			
	Insurance: Fire — Liability			
	Building Permit			
	Temporary Service Water & Lights			
PREPARATION OF LOT	Removing trees			
	Clearing site			
	Excavation			
	Hauling			
FOOTINGS	Excavation			
	Concrete footings			
	Backfilling			
MASONRY MATERIALS	Block			
	Common brick: Piers			
	Foundation			
	Chimney: Brick			
	Fire brick			
	Damper, ash dump, cleanout			
	Tile for hearth			
	Flue lining			
	Face brick			
	Cleaning brick			
	Sand			
	Brixment			
	Waterproofing (where used)			
	Drain tile (where used)			
	Back filling (basement)			
	Scaffold			
STEEL	Lintels			
	Reinforcing rods and mesh			
	Steel sash or vents			
	Pipe piers and columns			
	Flitch plates, bolts, and anchors			
MASONRY LABOR	Skilled			
	Common			
CONCRETE WORK	Basement floor			
	Porches			
	Stoops			
	Steps			
	Walks .... Drives			
	Garage or carport floor			
	Forms			
ROUGH LUMBER	Joist, girders and wall plates			
	Stud, plates and purlins			
	Rafters, ridge and valleys			
	Sheathing and Sub-floors			
	Bracing and headers			
	Stringers, ties, and bridging			
ROOFING	Shingles			
	Building paper			
	Flashing chimney and eaves			
ROUGH HARDWARE	Nails			
	Building paper			
	Bolts, anchors and ties			
EXTERIOR MILLWORK	Front doorway			
	Doors and frames			
	Windows and frames			
	Porch columns and rails			
	Moulds			
	Louvres			
	Shutters			
	Garage Door			
FINISH LUMBER	Porch and carport ceiling			
	Porch and carport box			
	Soffit and fascias			
	Moulds			
	Siding or shingles			
SCREENS	Doors .... Windows			
	CARRY FORWARD ☞			

			Estimated	Actual
	BROUGHT FORWARD ☞		$	$
STORM SASH	Doors .... Windows			
INTERIOR MILLWORK	Doors and frames			
	Interior casings			
	Mantel			
	Flooring (Hardwood)			
	Moulds, base and shelving			
	Stairway (where shown)			
KITCHEN CABINETS	Shop built			
	Built on job			
CARPENTER LABOR	Rough			
	Finish			
SHEET METAL WORK	Doors and Window Flashing			
	Termite Shields			
	Gutters and Downspouts			
	Splash blocks			
INTERIOR WALL FINISH	Lath and Plaster yds......... @........			
	Sheet-rock sq. feet........... @...........			
	Special Dry Wall finish sq. feet.......			
	Corner bead			
FINISH HARDWARE	Locks ....Hinges			
	Cabinet hardware			
	Screen hardware			
WIRING	Electric openings			
	Fixtures and installation			
	Dishwasher ......Dryer			
	Hot water heater ....Bell			
	Furnace .... Washer			
	Range ..... Air Conditioning			
PAINTING AND PAPERING	Caulking			
	Exterior woodwork			
	Interior woodwork			
	Interior walls and ceiling			
	Kitchen ....Bath			
FLOORS	Sanding and finishing (hardwood)			
SPECIAL WALLS AND FLOORS	Kitchen wall........... floor........... ...			
	Bath wall .... floor..................			
	Other			
PLUMBING	Cutting street and ditches			
	Septic tank........ Well & Pump.......			
	Fixtures .... and piping...........			
	Hot water heater			
	Medicine Cabinets			
	Towel, paper, and soap racks			
	Labor			
HEATING	Unit .... Tank..................			
	Ducts			
AIR CONDITIONING	Unit .... Installation..............			
ORNAMENTAL IRON	Columns .... Rails			
INSULATING	Walls .... ceilings .... floors.......			
WEATHER STRIPPING	Windows .... Doors			
CLEANING AND HAULING	Windows .... House..............			
	....			
GRADING, SEEDING AND PLANTING	....			
	....			
SHADES OR BLINDS	....			
SPECIAL EQUIPMENT	Refrigerator.......... Dishwasher.........			
	Range top & Oven .... Washer........			
	Exhause Fan........... Dryer............			
	Hood & duct .... Disposal...........			
OTHER ITEMS	....			
	Labor and Materials (Total)			
	Taxes and Insurance			
	Contractor's Fee			
	Total Cost			

ESTIMATING TABLES

BOARD FEET OF LUMBER

2 x 4

Half for 1 x 4 — Double for 2 x 8

Uo. Pcs.	10'	12'	14'	16'	18'
1	6⅔	8	9⅓	10⅔	12
2	13⅓	16	18⅔	21⅓	24
3	20	24	28	32	36
4	26⅔	32	37⅓	42⅔	48
5	33⅓	40	46⅔	53⅓	60
6	40	48	56	64	72
7	46⅔	56	65⅓	74⅔	84
8	53⅓	64	74⅔	85⅓	96
9	60	72	84	96	108
10	66⅔	80	93⅓	106⅔	120
20	133⅓	160	186⅔	213⅓	240
30	200	240	280	320	360
40	266⅔	320	373⅓	426⅔	480
50	333⅓	400	466⅔	533⅓	600

2 x 6

Half for 1 x 6 — Double for 2 x 12

No. Pcs.	10'	12'	14'	16'	18'
1	10	12	14	16	18
2	20	24	28	32	36
3	30	36	42	48	54
4	40	48	56	64	72
5	50	60	70	80	90
6	60	72	84	96	108
7	70	84	98	112	126
8	80	96	112	128	144
9	90	108	126	144	162
10	100	120	140	160	180
20	200	240	280	320	360
30	300	360	420	480	540
40	400	480	560	640	720
50	500	600	700	800	900

2 x 10

No. Pcs.	10'	12'	14'	16'	18'
1	16⅔	20	23⅓	26⅔	30
2	33⅓	40	46⅔	53⅓	60
3	50	60	70	80	90
4	66⅔	80	93⅓	106⅔	120
5	83⅓	100	116⅔	133⅓	150
6	100	120	140	160	180
7	116⅔	140	163⅓	186⅔	210
8	133⅓	160	186⅔	213⅓	240
9	150	180	210	240	270
10	166⅔	200	233⅓	266⅔	300
20	333⅓	400	466⅔	533⅓	600
30	500	600	700	800	900
40	666⅔	800	933⅓	1066⅔	1200
50	833⅓	1000	1166⅔	1333⅓	1500

For lengths over 18'0" Double 10', 12', or 14' Lengths

NAILS

Use	lbs.	Size	Material
Built-up girders	10	20d Com	100 lf
Floor joist framing	30	20d Com	1M bf
2x6 Ceiling joist-rafters	15	16d Com	1M bf
Frame walls—partitions	10	16d Com	1M bf
Frame walls—partitions	5	10d Com	1M bf
1" Sheathing—sub floor	20	8d Com	1M bf
1" Plaster grounds	10	6d Com	1M lf
1x3" Cross bridging	4	10d Com	100 lf
Bevel siding, wood	20	8d Cas	1M sq ft
T&G beaded ceiling	20	6d Cas	1M sq ft
⅞" Hardwood flooring	50	8d Cut	1M sq ft
Outside trim—moulds	20	8d Cas	1M lf
Outside trim—moulds	5	6d Cas	1M lf
Inside trim—moulds	15	8d Fin	1M lf
Inside trim—moulds	10	6d Fin	1M lf
Asphalt strip shingles	4	1" Gal	100 sq ft
Lath, gypsum	10	1⅛" Lath	1M sq ft

INTERIOR FINISH

LINE A Perimeter of room in lineal feet—Use for base and shoe.
LINE B Square feet in room (Floor or Ceiling)—Use for flooring, ceiling.
LINE C Square feet of wall and ceiling combined for eight foot ceiling. Use for figuring Plaster, Drywall, Painting—For exact amount of material deduct window and door openings. For Walls ONLY deduct amount shown in line. B.

Room Size Feet		2'	3'	4'	5'	6'	7'	8'	9'	10'	11'	12'	13'	14'	15'
	A	8	10	12	14	16	18	20	22	24	26	28	30	32	34
2'	B	4	6	8	10	12	14	16	18	20	22	24	26	28	30
	C	68	86	104	122	140	158	176	194	212	230	248	266	284	302
	A	10	12	14	16	18	20	22	24	26	28	30	32	34	36
3'	B	6	9	12	15	18	21	24	27	30	33	36	39	42	45
	C	86	105	124	143	162	181	200	219	238	257	276	295	314	333
	A	12	14	16	18	20	22	24	26	28	30	32	34	36	38
4'	B	8	12	16	20	24	28	32	36	40	44	48	52	56	60
	C	104	124	144	164	184	204	224	244	264	284	304	324	344	364
	A	14	16	18	20	22	24	26	28	30	32	34	36	38	40
5'	B	10	15	20	25	30	35	40	45	50	55	60	65	70	75
	C	122	143	164	185	206	227	248	269	290	311	332	353	374	395
	A	16	18	20	22	24	26	28	30	32	34	36	38	40	42
6'	B	12	18	24	30	36	42	43	54	60	66	72	78	84	90
	C	140	162	184	206	228	250	272	294	316	338	360	382	404	426
	A	18	20	22	24	26	28	30	32	34	36	38	40	42	44
7'	B	14	21	28	35	42	49	56	63	70	77	84	91	98	105
	C	158	181	204	227	250	273	296	319	342	365	388	411	434	457
	A	20	22	24	26	28	30	32	34	36	38	40	42	44	46
8'	B	16	24	32	40	48	56	64	72	80	88	96	104	112	120
	C	176	200	224	248	272	296	320	344	368	392	416	440	464	488
	A	22	24	26	28	30	32	34	36	38	40	42	44	46	48
9'	B	18	27	36	45	54	63	72	81	90	99	108	117	126	135
	C	194	219	244	269	294	319	344	369	394	419	444	469	494	519
	A	24	26	28	30	32	34	36	38	40	42	44	46	48	50
10'	B	20	30	40	50	60	70	80	90	100	110	120	130	140	150
	C	212	238	264	290	316	342	368	394	420	446	472	498	524	550
	A	26	28	30	32	34	36	38	40	42	44	46	48	50	52
11'	B	22	33	44	55	66	77	88	99	110	121	132	143	154	165
	C	230	257	284	311	338	365	392	419	446	473	500	527	554	581
	A	28	30	32	34	36	38	40	42	44	46	48	50	52	54
12'	B	24	36	48	60	72	84	96	108	120	132	144	156	168	180
	C	248	276	304	332	360	388	416	444	472	500	528	556	584	612
	A	30	32	34	36	38	40	42	44	46	48	50	52	54	56
13'	B	26	39	52	65	78	91	104	127	130	143	156	169	182	195
	C	266	295	324	353	382	411	440	469	498	527	556	585	614	643
	A	32	34	36	38	40	42	44	46	48	50	52	54	56	58
14'	B	28	42	56	70	84	98	112	126	140	154	168	182	196	210
	C	284	314	344	374	404	434	464	494	524	554	584	614	644	674
	A	34	36	38	40	42	44	46	48	50	52	54	56	58	60
15'	B	30	45	60	75	90	105	120	135	150	165	180	195	210	225
	C	302	333	364	395	426	457	488	519	550	581	612	643	674	705
	A	36	38	40	42	44	46	48	50	52	54	56	58	60	62
16'	B	32	48	64	80	96	112	128	144	160	176	192	208	224	240
	C	320	352	384	416	448	480	512	544	576	608	640	672	704	736
	A	38	40	42	44	46	48	50	52	54	56	58	60	62	64
17'	B	34	51	68	85	102	119	136	153	170	187	204	221	238	255
	C	338	371	404	437	470	503	536	569	602	635	668	701	734	767
	A	40	42	44	46	48	50	52	54	56	58	60	62	64	66
18'	B	36	54	72	90	108	126	144	162	180	198	216	234	252	270
	C	356	390	424	458	492	526	560	594	628	662	696	730	764	798
	A	42	44	46	48	50	52	54	56	58	60	62	64	66	68
19'	B	38	57	76	95	114	133	153	171	190	209	228	247	266	285
	C	374	409	444	479	514	549	584	619	654	689	724	759	794	829
	A	44	46	48	50	52	54	56	58	60	62	64	66	68	70
20'	B	40	60	80	100	120	140	160	180	200	220	740	260	280	300
	C	392	428	464	500	536	572	608	644	680	716	752	788	824	860

required to apply the waterproofing material? Will you do the work yourself or hire someone to do it for you? A call to your local builders' supply for the cost of materials may reveal it would be cheaper to have someone do the complete job for you. Call a waterproofing contractor for prices, and compare them to the material and labor cost. Then put the best price on your estimate sheet.

Backfilling the foundation walls is the item that must be estimated next. Calls to local excavating companies will give you the cost per hour of the heavy equipment you need to hire for this job.

Next you should estimate the cost of building the deck, or concrete floor, on which the house sits. Based on your plans, make a list of all the rough lumber material you will need to build the deck. Call your local lumber or builders' supply for their cost. Don't forget to price your nails, bolts, anchors, and ties. This will be your estimated material cost.

Will you do the work yourself or will you need help? You need some help for this heavy job, so estimate your labor cost according to the size of your house and the number of helpers you need.

The rough framing estimate will require listing the rough lumber needed for the exterior walls, interior walls, roof trusses or joists, and plywood sheeting for the roof. Consult your plans.

Determine the lumber needed, and call your lumber or builders' supply for the cost. What is the cost difference between buying trusses and building ceiling joists? Consider the amount of labor that it takes to build ceiling joists compared to putting up trusses.

Also check out the possibility of subcontracting all the rough framing. Compare the subcontracting cost to your material and labor cost, and determine which would be the best for you. When you make your decision, write the amount on your estimate sheet.

After you price the rough lumber, you need to determine how much roofing your building plans call for, and get a quote price from your building material supplier. Also check the cost of hiring someone to put the shingles on the roof for you. What would it cost if you do the work yourself? How long will it take? Time is important, and a roofing contractor may cost only slightly more than the cost of the material. Compare the cost of material and labor to contracting the work. Write down the best price, and this will be your estimated cost for applying the roof.

Windows are very important to your new house, and your plans will help you determine what kind of windows you choose. A double-glazed, or thermal pane, wood window offers excellent insulation. Aluminum windows with storm sashes installed also have good insulating properties. Whether your plans call for casement, double-hung, or awning-type windows, it is best to buy them pre-hung. Shop around by getting different building material suppliers to give you the best price. You will find there is great variation in window prices. The building materials supplier will give you different booklets containing information on the windows he sells. The best quality windows will not necessarily be the most expensive. Determine which you want in your home, and write the amount on your estimate sheet.

The sub-siding of the exterior walls is your next estimating job. The sub-siding usually contains some type of insulating material, and there are several materials available from your local supplier. The salesman will tell you that styrofoam in 4 x 8 or 4 x 12 sheets will give you the best "R" factor for insulating purposes. Aluminum-backed press paper is also available. Celotex is an inexpensive sub-siding which is effective and also has insulating properties, due to its being asphalt soaked.

Maybe your plans call for wood sub-siding. These may be in the form of 1′ x 6′ or 1′ x 8′ pieces of lumber put on your house in a diagonal position. This is an expensive way of sub-siding your home. Contact various building material suppliers and lumber companies for the price of the material you need. Then list the price you have chosen on your estimate sheet.

The exterior doors can be estimated next. Many do-it-yourself stores have a good selection of doors at a reasonable price, but your best price will probably come from the lumber company or the builders' supply. However, you may find the do-it-yourself stores carry the same door that the lumber company and builders' supply does, but at a lower price. What is the best price for you? Pre-hung steel doors are best for insulating and come in a wide variety of designs. Your plans will determine what doors you choose, but keep in mind that steel doors can be substituted for wood doors in many home designs without any noticeable difference. As a homeowner you will notice the insulating difference with the steel. Do your comparative shopping, and write your figures down on your estimating sheet.

Your outside trim should be estimated next. The outside trim of your house plans may determine that you will need a variety of items including 4′ x 8′ plywood good on one side, fascia board, nails, vents, and roof edging. Shop around by calling various building material suppliers and lumber companies for their prices. The amount you need depends on your house plans. Often these items do not appear on the list of materials supplied with your blueprints, but don't forget them. Once you have determined what you need and a satisfactory price, write the amount down on your estimate sheets.

The cost of your rough electric wiring is determined by two factors. Will you do the work yourself or hire it done? If you are going to hire an electrical contractor, you need to contact several different contractors, and let them give you an estimate from your plans. If you plan to do the electrical work yourself, you need quotes on the material you need. Electrical suppliers usually give you a good discount, and in some cases, you may find electrical wiring at discount department stores more reasonable than the electrical supply. Many discount department stores have a complete electrical department. Compare several of these stores' prices with the prices of electrical supply stores. Select the best price for you, and enter the amount on your estimate sheet.

Figure the estimated cost of your rough plumbing next. Your plans will determine what plastic pipe you need. How much it will cost depends on the supplier who gives you the best quoted price. When you have called different plumbing suppliers and determined the best price, then call several plumbers for an estimate. Now decide whether you will install your own plumbing or have a professional tradesman do the work. When you have considered the price of the material and your time involved compared to the plumber's quoted price, write the amount you feel best suits you on your estimate sheet.

Will you have a septic tank? Does your building lot have sewer service at the street, or has sewer already been stubbed onto your building lot? If you need a septic tank or sewer hook-up, now is the time to figure your cost. If you need only a sewer hook-up from the street, the best way to estimate your cost is by calling several local excavating firms that do this type of work. You will find that their price doing the whole job for you could be much cheaper than the cost of buying the material, renting the equipment, and hiring an equipment operator to do the job.

You may also need to know if you must post a bond to have the existing street torn up to bring sewer to your building lot. A call to your local municipal authority will give you the location of the sewer on your street and inform you of the cost of the bond, if any is required. This bond cost may vary, depending on the location of the sewer line and the amount of street or road damage necessary to bring the sewer service onto your building lot.

When a septic tank is necessary, it is important to determine how much leach line you need and the size of your septic tank. Call local excavating firms for a price to install the complete septic tank service, or call septic tank installers for a quoted price. Then get the cost of the leach line from your building material supplier, and the cost of digging the leach line trench and hole for the septic tank from several local excavating companies. Equipment can be rented with an operator on a per-hour basis.

Concrete block companies usually build septic tanks. Call them for a price on a septic tank. Sometimes burial vault companies also build septic tanks. Call several companies for the best price.

When you have gathered all your figures for a comparison list, you will find it may be more economical if you do all the contracting yourself. When you decide what is best for you, write your figure down on your estimate sheet.

What about your water service? Will you need a well or is there public utility water available on your lot? Call a local excavator for an estimate of a water hook-up service. If you need a well, you

This form will help keep your list of subcontractors clear since every job is itemized. A list like this one will probably be required by your lending institution.

The following is a true and complete list of subcontractors and those furnishing material for the construction of a building or improvements for ______________________________

on ______________________St., or Lot No. ____________________Montgomery County.

CONTRACT	CONTRACTOR	ADDRESS	AMOUNT		
Excavating					
Footings—Foundation					
Drain Tile—Water Proofing					
Beams—Columns					
Cement—Gravel					
Lumber					
Rough Carpenter Labor					
Brick—Brick Work					
Steel					
Stucco—Stone					
Fireplace—Flues					
Sheet Metal					
Roofing					
Wiring					
Plumbing					
Insulation					
Lathing—Plastering					
Steel Sash					
Mill Work—Cabinets					
Finish Carpenter Labor					
Hardware					
Floor Finishing					
Heating					
Softener—Cistern					
Pump—Well					
Septic Tank					
Electric Fixtures					
Ornamental Iron					
Screens					
Storm Sash					
Linoleum					
Tile					
Decorating					
Painting					
Walks and Drive					
Sodding					
Landscaping					
Garage					
Permits					
Land					
TOTAL					

STATE OF OHIO, COUNTY OF MONTGOMERY, ss:

-- being sworn says that he is the original contractor for the construction of certain building improvements for--on real estate above described and that the foregoing is in all respects a true, full and complete list of all subcontractors and of every person furnishing machinery, material or fuel towards the construction of said improvements and shows in full the amounts due or to become due to them for the performance of their respective contracts.

Sworn to before me by the said--and by him signed in my presence this

--------------------------19-----

(SEAL)

Notary Public, Montgomery County, Ohio.

SPEC SHEET FOR NEW HOUSES AND/OR CONSTRUCTION LOANS

ADDRESS ______ Number ______ Street ______ City ______ Twp. ______ County

LOT ______ Number ______ Plat ______ Size ______ Dimensions

DIRECTIONS ______ Specify how to locate

DESCRIPTION OF BUILDING: ______

Slab ☐ Bsmt. ☐ Crawl ☐ ______

HEAT ______ Type of fuel, description of system

FOR MULTI-LEVEL ONLY
Terrace Level is to be
Finished ☐ Left Unfin. ☐

FOUNDATION ☐ Conc. Blk. ☐ Poured Conc.

AIR CONDITIONING ☐ Central ☐ Wall Units ☐ How Many ☐ Air Purifier ______ Type

FIREPLACES ☐ Number ☐ Brick ☐ Stone ☐ Other

BUILT-INS ☐ Separate Oven ☐ Range Top ☐ Dropin Range Combination ☐ Disposal ☐ Hood

☐ Kit. Vent. Fan ☐ Dishwasher ☐ Food Center ☐ Refrigerator (Built-In) ☐ Inter-Com

Other ☐ Describe ______

NUMBER OF BATHS ☐ Full ☐ 1/2 Bath (2 fix) ☐ Add'l Lav. ☐ Add'l Stools ☐ Showers

BATH ONE
WALLS: Ceramic ☐ Plastic ☐ Metal ☐
FLOORS: Ceramic ☐ Vinyl ☐ Asp. Tile ☐
RECESS: Ceramic ☐ Other ☐

BATH TWO
Ceramic ☐ Plastic ☐ Metal ☐
Ceramic ☐ Vinyl ☐ Asp. Tile ☐
Ceramic ☐ Other ☐

STORM WINDOWS Storm Windows ☐ Storm Doors ☐ DBL Glazed Windows ☐ DBL Glazed Windows ☐

INSULATION Walls ☐ Ceiling ☐ Other: ______ Specify

WINDOWS: TYPE: ______ (Casement, awning, DBL-H, etc) ☐ Wood ☐ Alum. ☐ Steel

WOODWORK: (Jambs, trim) ☐ Pine ☐ Birch ☐ Mahogany ☐ Other ______ Specify

FLOORS: ☐ Oak ☐ Plywood ☐ Asp. Tile ☐ Vinyl Tile ☐ Asbest. Tile Other: ______ Specify

WALLS: ☐ Plaster ☐ Drywall ☐ Other ______ Specify

DRIVEWAY: ☐ Concrete ☐ Blacktop ☐ Other ______ Specify

YARD: ☐ Sod ☐ Seed ☐ Shrubbery ☐ Trees

Describe any other items of particular that you feel may have a bearing on the total cost of construction, either increasing or decreasing final cost. ______

______

SPECIFICATION SHEET FOR NEW CONSTRUCTION

ADDRESS ____________________

LOT ____________________

DESCRIPTION OF BUILDING ____________________

SITE IMPROVEMENTS: Sanitary Sewer ___ Septic ___ Public Water ___ Well ___ Cistern ___

FOR TRI-LEVELS, BI-LEVELS, BASEMENTS, LOWER LEVEL
FINISHED Ceiling: Tile ___ Drywall ___ Dropped ___ Walls: Drywall ___ Paneling ___ Floor: Carpet ___ Tile ___
UNFINISHED ___

FOUNDATION: Conc. Blk. ___ Slab ___ Poured Conc. ___ Bsmt. ___ Crawl ___

HEATING AND COOLING SYSTEM: ____________________
Type of Heating System
Type of Fuel: Electric ___ Oil ___ Gas ___ Other ___

Central Air ___
Electric Air Filter ___
Humidifier ___
Capacity of A/C ___ tons

WINDOWS: ____________________
Casement, Dbl.-H., etc.
Wood ___ Alum ___ Other ___

Storms ___
Insulated: Dbl. Glaze ___ Trpl. Glaze ___
Thermal Break ___

FRAMING: ____________________
Method & Materials

ROOFING SYSTEM: ____________________
Method & Materials

INSULATION: Walls ___ Ceiling ___
R-factor R-factor
____
Type of Insulation

Type of Sheathing: Fiberboard ___ Styrofoam ___ Other ___

INTERIOR WALLS: Drywall ___ Plaster ___ Other ___

EXTERIOR FINISH: ____________________

INTERIOR TRIM: Pine ___ Other ___

Stained ___ Painted ___ Cabinets (type) ___ Counter Tops ___

INTERIOR DOORS: Smooth Hollow Core ___

Panel ___

LIGHT FIXTURES: ALLOWANCE$ ___

Special Light Fixtures ___

FLOOR COVERING: ALLOWANCE$ ___

Carpet ___ Other ___
Tile ___ Linoleum ___ Hardwood ___
Special Entries, etc. ___

BATHS: Full ___ Half ___ Add'l. Lav. ___

BATHROOM FINISH:
Walls: Ceramic ___ Fiberglass ___ Other ___
Floors: Ceramic ___ Vinyl Tile ___ Linoleum ___ Carpet ___

FIREPLACES: Number ___

Masonry Const. ___ Metal Prefab. ___ Heat-a-lator ___
Brick Finish ___ Stone Finish ___ Other ___

BUILT-INS AND OTHER EXTRAS: Drop-in Range ___ Range Top ___ Separate Oven ___ Micro Range ___

Dishwasher ___ Disposal ___ Hood & Fan ___ Compactor ___
Intercom ___ Softner ___ Gar. Opener ___ Other ___

DRIVEWAY: Concrete ___ Blacktop ___ Other ___

YARD: ALLOWANCE$ ___ Sod ___ Seed ___ Shrubbery ___ Trees ___

SPECIAL FEATURES (Cathedral Ceilings, Sunken Living Room, Special Elevations, etc.)

Builder's Signature ____________________
Date ____________________

12/77 Rev.

Two types of specification sheets used by lenders.

need to get quotes from different well drilling firms as to how much their charge is. The well-digger usually quotes you a price per foot for a well including casing. He also expects to sell you the pump for your well and will take the complete water service to just inside your new home. You are usually required to provide an open trench from the well to the building and the cost of covering the water lines once they have been approved. When you have well-diggers, estimate the pump, well, and insulation, and write it down on your estimate sheet. Also include the cost of digging the open trench from the well to the building and the cost of covering the water lines once they are in.

The interior walls are estimated next, and the information you need is on your plans. Walls are usually of gypsum board, but sometimes plaster is used. The most economical material is gypsum board, also known as drywall or sheetrock. The boards are in 4′ x 12′ or 4′ x 8′ sheets and must be finished before your walls can be painted. Therefore you must estimate the cost of the gypsum board by calling local builders' suppliers and lumber companies to get your best price. You also need to include the cost of labor to hang the gypsum board or drywall if you don't do the work yourself.

Material to finish your sheetrock walls after they have been hung is also necessary. You need to call your local building suppliers for the best price of tape, all-purpose joint compound, topping compound, and corner beads which are used to make your walls ready for painting. Finishing your drywall is called spackling, and this should be done by a professional. Call several drywall finishers listed in the phone directory, and ask them to quote you a price per square foot for finishing your walls. Don't let the spackling contractor talk you into letting him supply the material and do the hanging and finishing of your walls. If you do, the price will be 20 to 25 percent higher for the complete job.

When you have selected a price from the spackling contractor, write all the cost of finishing your interior walls on the estimate sheet. Unless you are a professional drywall finisher, it would be in your best interest to contract the finishing of your walls.

Your house plans may have certain areas where the walls will have an added cost, such as the kitchen and bathrooms. Ceramic and plastic are the most common types of walls for the kitchen and the bathrooms. The best place to get a price on these materials is at stores that specialize in tile and vinyl wall or floor covering. Visit some of these stores to see what is available and get their cost. Once you have decided on a material and price, include this figure on your estimate sheet.

When you have estimated the cost of the interior walls, you will need paint to paint them. Off-white is a preferable base paint to use. Visit your local paint stores, and shop for the lowest cost, water-base interior paint. Have the salesman or store manager quote you a price per 5-gallon bucket. You will probably need several buckets, so you should ask for and get a good discount. Be sure to deal with reliable paint stores.

Write the cost of the amount of paint you need for the interior walls on your estimate sheet. Also include the labor, if any, of putting the paint on the walls. Don't contract a painter to paint the interior of your home. This is a simple job and can be done without the expense of a painter.

Count the doors and frames you need. Figure the amount of interior casings, hardwood flooring, moldings, vase, shelving, and a mantel if you need one. Interior doors are best purchased prehung and are easily installed. Your local building supplier or lumber company will have brochures of different interior doors and trim. Select those with the best price and the ones that meet the needs of your house plans. Enter the amount on your estimate sheet.

Kitchen cabinets and bathroom vanities should be figured next. Take your building plans to several different cabinet and kitchen specialists for a price. The kitchen specialist will help you decide what kitchen cabinets and bathroom vanities you need. Some cabinet and kitchen plan-

For a difference of only a few hundred dollars on a typical $50,000 home, you can have 2 x 6 studs in the walls of your home rather than the traditional 2 x 4, shown here leaning against its big brother for comparison. The extra two inches allow for extra insulation later in the building process.

ning specialists give a builder's discount of up to 50 percent off the cost of the kitchen cabinets and bath vanities, so do some comparative shopping for your cabinets.

Tops for the cabinets could be selected at the time you select your cabinets. There is usually no discount allowed for the cabinet tops. Installation of vanities and kitchen cabinets is not difficult, and if you do the work yourself, no labor will be required. When you have chosen the kitchen cabinets, vanities, and tops you want, write their cost on your estimate sheet.

Kitchen appliances, which are often considered special equipment, need to be estimated next. Ask your kitchen specialist if he also supplies appliances. If he does, buy from him because you will probably get a bigger discount. If he doesn't, shop several discount stores for a complete appliance package. Then write the cost on your estimate sheet.

Will you stain or paint your woodwork? Now is the time to get a price for the interior trim. Just select what your plans call for and contact your local paint stores for the best price of oil-base paint or wood stain. Enter the amount on your estimate sheet.

When estimating your finished exterior walls, it is best to keep in mind the different discounts one masonry supplier might offer over another. However, when you call for a quote on the price of brick per thousand, don't write the discounted price on your estimate sheet as you may not have the money to pay the supplier by the specified discount date. Determine the brick you need along with the amount of mortar, wall ties, and sand. Once you have the price of brick, mortar, etc., from your brick supplier, write the amount on your estimate sheet. Then contact several bricklayers, usually listed in the telephone directory as masonry contractors. When you contact the one you feel has the best price, write the estimate down on your estimate sheet.

Pre-hung doors, like this one awaiting installation, eliminate the inconvenience and possibility of error in hanging a door on hinges. The entire frame will fit into the doorway, its position adjusted with wooden wedges as shims. Spaces around a pre-hung door, as with a custom one, are concealed under molding strips.

To get an estimate for your downspouting, just call several spouting contractors, who are listed in the phone book. This is a job which should be subcontracted. If you want to do the work yourself, do it. The material may cost you as much as the subcontractor would charge for the material and labor combined. This is because the subcontractor often buys in such large quantities that his prices are difficult to beat.

Call your local builders' supply for the cost of splash blocks. Once you have your cost figures for guttering, downspouts, and splash blocks, write them down on your estimate sheet.

Plumbing fixtures can be estimated next. Contact your local plumbers' supply for information about the latest designs in plumbing fixtures. The plumbers' supply store has several pamphlets and booklets from which you may select the fixtures you prefer. They will also supply you with a price list.

Haggle over prices with the salesman. You can get a good price if you simply ask for a discount. Some places don't juggle prices, but many do. Write the fixtures and the prices on your list, add them together, and enter the amount on your estimate sheet.

You also need to include the cost of faucets and traps with your estimate figures. These can be acquired when you select your fixtures. Don't forget solder, paste, and sandpaper.

You do not need to estimate your electrical finish if you subcontract the electrical work. However, if you plan to do the wiring yourself, you need to estimate the cost. Prices of switches, switch boxes, cover plates, outlet boxes, and light fixtures such as lamps, fans, intercom, or other items should be included in your finished electrical cost. Talk with the salesmen at different electrical supply stores and ask for a builders' discount. Don't buy from the electrical supply stores unless they agree to give you a discount.

Go to a discount department store, and check the prices of the material you need. A discount store's prices will generally be the same as the electrical supply prices with the discount you demand.

Shop very carefully for your light fixtures which are very often overpriced in electrical sup-

ply stores. Write the cost on your estimate sheet. If you are subcontracting the work, just enter the figures the electrical contractor gives you.

Cost of your porches, stoops, steps, walks, garage floor, and patios should be estimated next. Measure the length, width, and depth of each to figure the amount of rough lumber you need to build these forms. You already have the cost per foot of rough lumber that was used in estimating your rough framing. Determine the cost of the forms which you need, and write the cost on your estimate sheet.

You also have your concrete prices per cubic yard. Determine how much concrete you need for walks, porches, stoops, and patios from your plans, and multiply by the cost. Write the amount on your estimate sheet.

Call different excavating contractors for the cost of labor and equipment for the final grade on your yard and around your house. Write the amount on your estimate sheet.

Will you seed or sod your lawn? Call your local sod farms for their price of sod per square yard. Also check with them about the cost of having them lay the sod for you. Determine how much seed and fertilizer you need if you plant grass yourself. Go to your local seed stores, and ask for their price of the seed and fertilizer you need. Then compare this cost to buying sod and laying it yourself, or buying sod and having the sod company lay the sod. Consider, too, the weeks you will have to spend, watering and weeding the seedlings. Select the cost which best suits your needs, and then write the amount on your estimate sheet.

Your interior floor cost is determined by the type, quality, and amount of floor covering you need. Hardwood is figured by the square foot and comes in a variety of prefinished stains. Call your local builders' supply and lumber company for the cost of tongue-and-groove prefinished hardwood per square foot. Multiply the cost by the amount of hardwood flooring your plans call for. Then write the amount down on your estimate sheet.

Labor cost to install the hardwood should also be included. The cost of a carpenter to put down your hardwood could be very expensive, but the job can be done by anyone who can follow instructions and use a heavy nail gun. Therefore, if you do the work, write down only the cost of the material.

If part or all of your floor covering will be carpet, shop in different discount carpet outlets for the best price. Shop all available sources you have time to contact for the best price. Carpet

(Left) Kitchen cabinets come in standard forms and sizes or can be made to individual specifications for very little extra cost. This open cabinet will house an increasingly common kitchen appliance — a double microwave oven ensemble.

(Right) When the cabinets arrive, they won't have handles. Do not panic. They are attached to the inside, for protection during shipping.

(Above) Before the carpet comes the pad, and it pays to get the best you can afford of both. The tack strips running along the wall bristle with short nail points. The carpet will be stretched and chiseled over these points and held in place by them. Another method of carpet installation is adhesive, usually done with carpet tiles, but nails are most satisfactory.

(Below) One reason you should rely on a professional carpetlayer to install your broadloom: specialized tools. This stretcher with extensions keeps the unattached carpet taut while it is being trimmed and tacked. If the carpet is not as tight as it can be, you may get billows and bubbles of loose rug fabric. The metal pole must reach all the way across the room where it is propped against the wall. What do they do in a ballroom? Use many, many extensions!

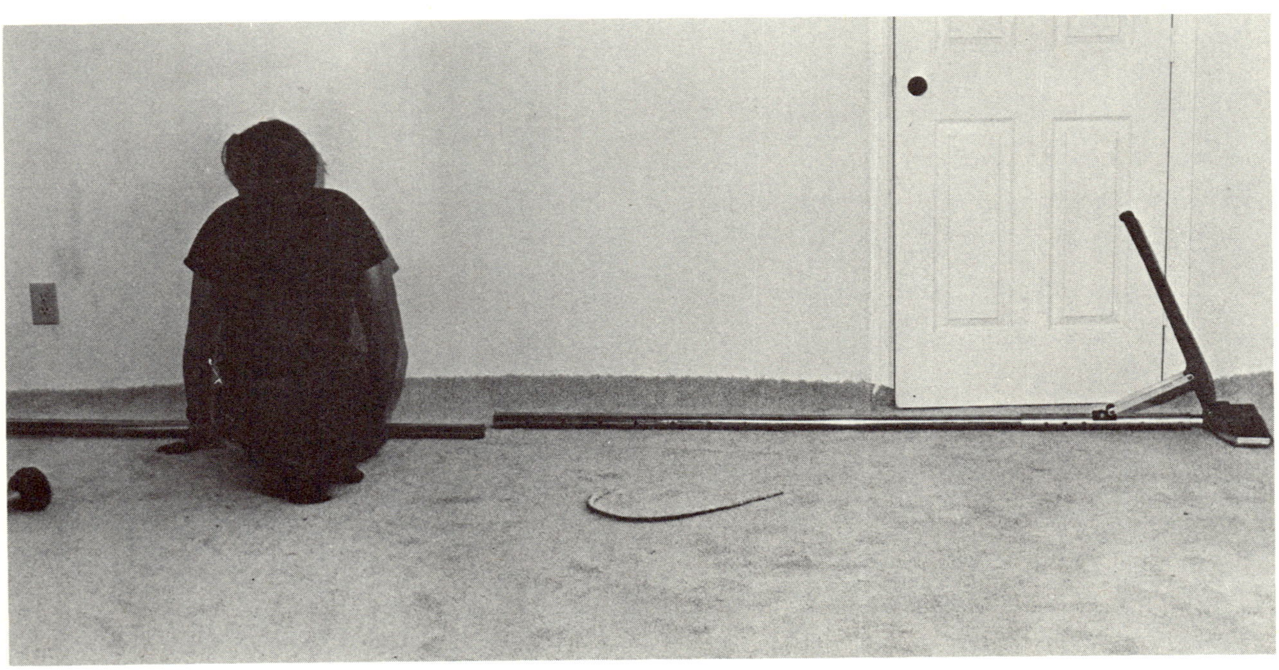

prices and quality are highly variable, and you could save several hundreds of dollars by shopping carefully. If you live near a carpet factory, be sure to check its outlet stores.

The best carpet pad is the waffle type, but batts of pressed foam rubber are available at a lower price. Choose the pad that best suits your wishes, and compare their cost when you are shopping for your carpet.

The labor cost for installing carpet is usually a set price in certain areas. If you know how to install carpet you may save money, but if you are not a professional carpet installer, you probably won't do a good job. Hire the professional as this is a finished area of your new home and, like the drywall, is always seen. When you are certain of the cost of the carpet you want and the cost of installing it, write it on your estimate sheet.

For areas where the floors are linoleum, vinyl, rubber or asphalt tile, such as the kitchen, bathroom, utility rooms, and entranceway, you could do the work yourself and probably get a professional looking job. The best place to buy floor tile and inlaid linoleum is a wholesale and retail store that specializes in wall and floor tile.

Remember that you are in the carpet or tile store only to shop for price and style. Do not be high pressured into buying any floor covering yet. Some carpet or tile stores will offer to hold your carpet for you, but if another customer wants the same product and has the cash, your carpet may go to someone else, and you will wind up waiting while yours is back-ordered.

Shop around for the best price. There is a wide variety of tile and linoleum to choose from. Although most reliable tile stores carry safe namebrands, remember to shop carefully. You need to include the cost of adhesive to lay the tile or linoleum. When you choose your material and its cost, enter the amount on your estimate sheet.

Once you have estimated all the items that are needed to build your house, add them all together, and put the total on the estimating form in the "Total Cost" column. When you have completed this form, examine it and your house plans to see if you have overlooked any items.

If you haven't overlooked anything, this total is the estimated cost of your house. These are figures you need when you go to the lending institution to secure your construction loan.

Another special tool used by carpetlayers is a super-hot iron, used to melt adhesive which joins carpet edges in an invisible seam. Several hundred degrees hotter than a household iron, this one glides along beneath the carpet and rests, when not in use, in its own metal holder (shown here immediately behind the tool box). If you buy or rent tools to lay your own carpet, USE THE HOLDER, or you will probably burn an expensive hole in the broadloom.

If you find one of these "stop work" orders on your house, do what it says — stop work and get your paperwork straightened out. The building department has ways to make life unpleasant for you. You may be ordered to tear down unpermitted work, your house may be condemned, or you may be cited into court.

County of Montgomery

DIVISION OF BUILDING REGULATIONS DEPARTMENT

451 W. THIRD ST. • PHONE 225-4622—225-4630

DAYTON, OHIO 45402

NOTICE TO STOP WORK. NO PERMIT

☐ BUILDING
☐ ELECTRIC
☐ HEATING

To Whom It May Concern:

Notice is hereby given that **NO PERMIT** has been issued____________________

for the construction, repair, alteration of, addition to, the____________________building

situated on Lot____________________Subdivision, Montgomery County, Ohio, and all work in, on or about such building must be suspended forthwith. Any person who shall continue the construction of said building, use any material in or about said building or use any machinery in or about said building, after the service or posting of this notice, will be subject to prosecution, pursuant to the provisions of The Montgomery County Building Code and the amendments thereof.

By Order of

____
Montgomery County Building Inspector

Date____________________, 19____ By____________________
Assistant Inspector

Chapter 6

Knowing Your Building Codes

It is absolutely necessary to know the building code, building regulations, or building rules in the area where you are building. There is no exception to this requirement. Ignoring building codes or working with inadequate information is no excuse. If you fail to conform to the building code in your area, your new home can be determined uninhabitable by the courts, and no one will be allowed to live there. This situation rarely occurs, but it can and does if you willfully neglect building codes.

However, most building regulation departments are more than willing to work with you and advise you according to their building regulations. A "County Building Code" book can usually be purchased from your county building regulation department. It is also possible your county has just a few regulations and doesn't issue books or booklets. Either way, you should get all the information available.

When you first acquire a building code or regulation book, you should familiarize yourself with the complete contents. If you have any questions, call your local building regulation department, and make an appointment to discuss any part of the codes or regulations that can't be explained by a phone call. Don't hesitate to take your plans to the building department for help and advice at any point of your planning or building.

Don't be like the homeowners who built their dream home in a Massachusetts woods. Their dream was a new house with no electrical service, no indoor plumbing, and no road or driveway. However, the nearby city and county building code said they must have all three. The couple were very energy-conscious and wanted to get back to nature. Their desire to have a beautiful home in an unpolluted woods has been blocked by court orders and denials of variance request. Although the couple's plans for their new home were approved by the county, and the house eventually was given a variance by the city, it now has to have a variance from the state to have a privy instead of a septic tank. Even if you want to be a pioneer, there are rules.

If the couple had known the building code for the area they had chosen, maybe they would have selected another area or state in which to build, but it is rather difficult to find an area that does not have strict building codes and regulations. To avoid having to argue for two or three years with a building regulatory agent, know your building codes and build your new home accordingly. The building regulations are for your own protection as well as to protect your environment. Even if you want to live without modern amenities or safe electricity, the local government has to think about future owners of the house.

Electrical codes in most areas are the national electrical code. Unless specific restrictions or regulations are written into the local code, you can assume the national electrical code is in effect in your area.

The plumbing code is usually a state code which is enforced through the local building regulation department. Different locations may add requirements to the state plumbing code as needed in special areas. Don't be surprised if you find some parts of the building code in your area are antiquated and do not take full advantage of modern technology in the building business.

The building regulation department works with the environmental health department in most areas. The environmental health department was established to protect our environment, but sometimes its members think their jobs are to harass the taxpayers. Beware of them. In some areas, anything that is more than 3 feet away from your building comes under their jurisdiction.

Unlike the building regulation department, electrical department, and plumbing department, the environmental health department sometimes thinks that everyone who wants to build a home is violating the environment. So be forewarned and deal with them very carefully. Once you know your building codes and whom you will be working with, take advantage of all the information you can get from them.

MONTGOMERY COUNTY
BUILDING REGULATION DEPARTMENT
County Government Plaza
451 W. Third Street

DAYTON, OHIO 45402 — **Telephone 225-4622**

Excerpt from Uniform Building Code of Montgomery County

Section 111 — Application for Permits.

Application shall be made by the owner of the building involved, or his authorized agent, architect, engineer or contractor employed in connection with the proposed work.

Applications shall be prepared in writing on printed forms prescribed by the County Building-Inspector* and shall contain the full names and addresses of owner, applicant and contractor, the street and house number of the building*, a brief description of the proposed work, the estimated cost thereof, a statement in behalf of the owner that the proposed work and occupancy* will be in accordance with all laws and regulations applying to the premises, and such other information as may be required by the County Building-Inspector*.

All applications for residential-building* permits shall be accompanied by two sets of plans drawn to scale, and specifications.

2 Plot Plans	2 Front Elevations
2 Foundation Plans	2 Side Elevations
2 Floor Plans	2 Rear Elevations
2 Wall Sections for House and Garage	

The plans shall be complete, showing all details and sections necessary to construct the building*, together with 2 copies of plot plan drawn to scale showing actual shape and dimensions of the lot to be built upon, the exact size and location on the lot of all existing* buildings*, the location of proposed buildings* with dimensions and distances to all lot lines, the existing and intended use of each building, the sanitary drainage and sewage disposal, the downspout, surface and cellar drains, water lines and source of water supply and any other information that may be necessary to determine compliance with this Code.

Where in the opinion of the County Building-Inspector* it is necessary, a survey plat in duplicate showing the surveyor's stakes shall also be provided.

The survey points shall be maintained and be accessible at all times during construction.

One set of such plans and specifications shall be returned to the owner when the permit is issued.

You should know the applicable building code requirements for your lot, and many regulation departments provide excerpts such as this one for handy reference. You should also request a copy of the entire applicable building code, too, for detailed research when needed.

Be familiar with the applications and permits required by the building department in your county, such as this application for a zoning certificate.

MONTGOMERY COUNTY BUILDING INSPECTION
COUNTY ADMINISTRATION BUILDING -- ROOM #918
451 WEST THIRD STREET DAYTON, OHIO
TELEPHONE 225-4622,-4623,-4626,-4627

APPLICATION FOR A ZONING CERTIFICATE

(To be Filled out by Applicant)

Applicant____Phone__________

Address____

____

The applicant is the owner, lessee, etc., of this property. Specify____________

Owners Name____

Address and Location of Property____________________________________

__________________________Township_____________Section__________

Said property contains approximately_____(acres) or (square feet), and is located along the north, south, east, west, (circle appropriate direction) side of______________

______________________________(road), (street), or (avenue), approximately__________feet north, south, east, west, (circle appropriate direction) of_______________________________________(road), (street), or (avenue), or Lot#_______and recorded plat name_________________________

Describe in detail, not generalities, how this property is to be used:_____________

____

____

____

____

Please attach a plan showing the following:

A. The actual dimension of the lot including easements.

B. The exact size and location of all buildings existing on the lot.

C. The proposed new construction.

D. The existing and intended use of all parts of the land or buildings.

Proposal

Morgan Electric Co.

PROPOSAL SUBMITTED TO	PHONE	DATE
Jim & Helen Brown		11-15-77
STREET	JOB NAME	
CITY, STATE AND ZIP CODE	JOB LOCATION: Fauber Road	
ARCHITECT	DATE OF PLANS	JOB PHONE

We hereby submit specifications and estimates for

Estimate to install electrical wiring.
Estimate to include:

400 Amp distribution breaker panel service with an underground entrance.
Extras, if any, will be billed on final statement.

- 165 Total wiring openings
- 47 Switches
- 69 Receptacles
- 45 Fixture openings
- 1 Hood wiring
- No Fixture allowance included
- 1 Fan in bath by office
- 1 Freezer circuit in utility room
- 1 100W Sq. recess over kitchen sink
- 1 Fan in half bath
- 1 Circuit to front of house for future post lights
- 1 Circuit to rear of house for future yard lights
- 1 60 Amp range circuit
- 1 30 Amp dryer circuit
- 1 40 Amp double oven circuit
- 1 30 Amp water heater circuit
- 2 60 Amp elec. furnace circuits
- 1 40 Amp air-cond. circuit
- 1 15 Amp disposal circuit
- 1 20 Amp dishwasher circuit
- 1 20 Amp washer circuit
- 1 Weather-proof G.F.I. circuit
- 1 Bathroom G.F.I. circuit
- 1 Complete door bell wiring
- Permit

Any changes or additions will appear on final statement.

We Propose hereby to furnish material and labor — complete in accordance with above specifications, for the sum of:

Two Thousand Four Hundred Fourty Three and no/100---------- dollars ($ 2443.00)

Payment to be made as follows:

65% Due on Rough---$1587.95

35% or Balance Due on Finish-----------------------------------$ 855.05

All material is guaranteed to be as specified. All work to be completed in a workmanlike manner according to standard practices. Any alteration or deviation from above specifications involving extra costs will be executed only upon written orders, and will become an extra charge over and above the estimate. All agreements contingent upon strikes, accidents or delays beyond our control. Owner to carry fire, tornado and other necessary insurance. Our workers are fully covered by Workmen's Compensation Insurance

Authorized Signature: Jerry Morgan

Note: This proposal may be withdrawn by us if not accepted within 30 days.

Acceptance of Proposal — The above prices, specifications and conditions are satisfactory and are hereby accepted. You are authorized to do the work as specified. Payment will be made as outlined above.

Signature ____________________

Date of Acceptance: ____________________

Signature ____________________

Chapter 7

Shopping for Subcontractors

Subcontractors may do a substantial part of the work on your new home. Shop for your subcontractors with the idea of getting the best and most work for your money. Quantity is important, but so is quality, so don't just look for fast workers. Get good ones.

You can look in a telephone directory and select a subcontractor, but if you do, it is the same as buying a "pig in a poke." The best way to hire subcontractors is to drive around an area where new homes are being constructed. Examine the work of the subcontractors. Stop and talk to different people on the job site. Make a note of the names of the subcontractors from the signs painted on their trucks.

Many subcontractors are one-man operations. Although the subcontractor may have several employees working for him, very often he runs his business out of the glove compartment of his truck. Therefore, call the small subcontractor in the evening. He probably maintains an office in his home and makes his calls in the early morning before he goes to work or in the evening when he has completed his day.

You will find most subcontractors are very cooperative and will suggest a fellow electrician, plumber, or drywall finisher. Ask him if he knows any who want extra work. If one subcontractor is too busy to help you, ask him if he knows anyone else to do the particular job you want done. The subcontractor can usually recommend someone with skill who needs work or who can work your job into his schedule.

A written proposal like this one, giving exact specifications and price, should be obtained from all subcontractors.

Determine what subcontractors you need for major jobs, such as your electrical work, drywall finishing, bricklaying, etc. Then determine the small subcontractors that you may use, such as a roofer, a concrete finisher, a gutter and downspout contractor, etc.

Unless you live in the desert and use no rain gutters, get a quote from a gutter and downspouting contractor. Then call several suppliers for the cost of guttering and downspouting. Figure the amount of time it will take you to install the gutters and downspouts. Get the cost of the extra tools you will need. Also include any labor cost other than yourself. When you add all your figures together, you may find it is cheaper to hire the guttering contractor to do the work than for you to do it yourself, buying materials separately.

Subcontractors who can do small jobs for you may save you money and time. The best way to determine if a small job contractor can save you money is to compare his cost in the same way as the cost of gutter and downspouting. Some small jobs for which you should consider using the subcontractor are: waterproofing, guttering and downspouting, roofing, concrete finishing, and termite pretreating.

Have each subcontractor supply you with an estimate of the cost of his job. He can gauge the work he will do for you from your plans and will then submit a contract for your approval. Study the subcontractor's proposal, and compare his prices with the quotes of at least one other subcontractor. Also be sure to compare the cost of do-

HEATING STEVENSON AIR CONDITIONING

SPFLD. 325-7011
DAYTON 864-7302
XENIA 426-3444

CONTINUOUS SPOUTING

H.S. STEVENSON & SONS, INC.

GENERAL OFFICES & SERVICE DEPT 2600 SOUTH LIMESTONE STREET
SPOUTING-GASOLINE-FUEL OIL DEPTS 2850 EAST LEFFEL LANE

P.O. BOX 1345
SPRINGFIELD, OHIO 45501

• AIR EASE
• CARRIER
• LENNOX

PROPOSAL NO. 2097

TO: Brown Construction

DATE August 24,1977

PHONE

CITY STATE ZIP

1. JOB: ADDRESS Fauber Road
2. EQUIPMENT: 1- Johnson/Lennox Oil Heat Pump System: 4 ton
3. ACCESSORY EQUIPMENT: Thermostat
4. PROPOSED WORK: All material and labor to completely install as per our layout.

*All Electrical wiring of equipment to be done by Electrician.

NOTE: 1- 550 Gallon Oil Tank (NOT BURIED) ADD: $ 310.00

OPTIONAL: 1- Lennox heat pump system: 4 ton

5. ADDITIONAL ITEMS (INDICATE IF APPLICABLE):

a. INSTALLATION & LABOR	Yes
b. PLUMBING	Drain for A/C
c. WIRING	Low Voltage only
d. MASONRY	Slab for Cond.Unit
e. CARPENTRY	Cut register openings
f. PERMITS	Not included
g. REMOVE & RETAIN EXISTING UNIT	None
h. MISCELLANEOUS	None

ITEMIZATION:

HEATING EQUIPMENT	$ 4,067.00
AIR CONDITIONER	
HUMIDIFIER	200.—
ELECTRONIC CLEANER	400.—
SPOUTING	
MISCELLANEOUS	3,670.00
MISCELLANEOUS	
TOTAL	

PROPOSAL

WE AGREE TO FURNISH & INSTALL IN A WORKMANLIKE MANNER THE ABOVE DESCRIBED LABOR & MATERIALS FOR $________, ON THE TERMS INDICATED BELOW. PRICES QUOTED HEREIN CANNOT BE GUARANTEED FOR MORE THAN 30 DAYS FROM ABOVE DATE.

FOR H.S. STEVENSON & SONS, INC. R. David Stevenson rdy DATE 8/24/77

ACCEPTANCE OF PROPOSAL

THE PURCHASER HEREBY ACCEPTS THE ABOVE PROPOSAL, SUBJECT TO THE CONDITIONS ON THE BACK AND TO THE FOLLOWING CONDITIONS, AND AGREES TO PAYMENT AS FOLLOWS: PARTIAL BILLINGS WILL BE MADE MONTHLY AS JOB PROGRESSES (UNLESS OTHER ARRANGEMENTS ARE SPECIFIED BELOW).

PAST DUE BALANCES (ANY BALANCE NOT PAID WITHIN 30 DAYS FROM DATE OF BILLING) WILL BEAR INTEREST AT THE RATE OF 1.5% PER MONTH OF THE UNPAID BALANCE, 18% ANNUAL INTEREST RATE.

PURCHASER X DATE 10-14-77

HEATING • AIR CONDITIONING • CONTINUOUS SPOUTING • ELECTRONIC CLEANERS • HUMIDIFIERS • GASOLINE • FUEL OIL

This proposal is for gutters and downspouts.

HEATING
STEVENSON
AIR CONDITIONING

SPFLD 325-7011
DAYTON 864-7302
XENIA 426-3444

CONTINUOUS SPOUTING

H.S. STEVENSON & SONS, INC.

GENERAL OFFICES: 2600 SOUTH LIMESTONE STREET
SPOUTING DEPARTMENT: 2850 EAST LEFFEL LANE
P.O. BOX 1345

• AIR EASE
• CARRIER
• LENNOX

TO: Brown Const Co
1121 Peoria Ave — Mailed 11/12/77 — DATE 10-6-77
Brentwood Ohio
CITY — STATE — ZIP — PHONE

PROPOSAL

WE HEREBY PROPOSE TO INSTALL THE FOLLOWING LISTED ITEMS OF GUTTER AND/OR DOWNSPOUT IN A WORKMANLIKE MANNER AT: ADDRESS Same CITY ______

White ALUMINUM GUTTER	267' FEET (APPROX.)	SPLASH BLOCKS	None
DOWNSPOUTS	3-3" 2-4"	ONE STORY/TWO STORY	One & Two
ELBOWS	9-3" 6-4"	UNDERGROUND DRAIN 4" LINE	None
MITERS	7-out	TEAR-OFF	No
FUNNELS	None	GUTTER SCREEN	None
HANGERS	None	—	—
—	—	—	—

WE AGREE TO FURNISH & INSTALL IN A WORKMANLIKE MANNER THE ABOVE DESCRIBED LABOR & MATERIALS FOR $ 320.05, ON THE TERMS INDICATED BELOW.

PRICES QUOTED HEREIN CANNOT BE GUARANTEED FOR MORE THAN 30 DAYS FROM ABOVE DATE.

Sammy Moore
FOR H.S. STEVENSON & SONS, INC.

ACCEPTANCE OF PROPOSAL

THE PURCHASER HEREBY ACCEPTS THE ABOVE PROPOSAL, SUBJECT TO THE CONDITIONS ON THE BACK AND TO THE FOLLOWING CONDITIONS, AND AGREES TO PAYMENT AS FOLLOWS:

CASH UPON COMPLETION OF WORK (UNLESS OTHER ARRANGEMENTS ARE SPECIFIED BELOW).

PAST DUE BALANCES (ANY BALANCE NOT PAID WITHIN 30 DAYS FROM DATE OF BILLING) WILL BEAR INTEREST AT THE RATE OF 1.5% PER MONTH OF THE UNPAID BALANCE, 18% ANNUAL INTEREST RATE.

11-14-77 DATE — x Herman Nelson Supervisor PURCHASER

HEATING • AIR CONDITIONING • CONTINUOUS SPOUTING • ELECTRONIC CLEANERS • HUMIDIFIERS • EXCAVATING

Another proposal, this one for an oil heat system.

Xenia, Ohio 45385
Phone 862-6711

James & Helen Brown

Date 11-15-77

DATE	CHARGES AND CREDITS		BALANCE
	BALANCE FORWARD		
	Contract	$2443.00	
	65% Due on Rough	$1587.95	
	35% or Bal. on Finish	$ 855.05	
	Rough Due		$ 1587 95

All accounts are due and payable by the 10th of the month following date of invoice.

Morgan Electric Co.

Thank You

PAY LAST AMOUNT IN THIS COLUMN

This electrical work subcontractor required 65 percent of the fee at the completion of the rough installation. Contracts should be as specific as possible.

ing the work yourself. If you determine that one particular subcontractor is best for you, sign the acceptance portion of both copies of the contract proposals and return one to the subcontractor. Be sure you note on your copy the date you mailed it back to the subcontractor.

Choose your subcontractors wisely by taking advantage of all available knowledge in reference to their work. Look at prior houses they have completed. Consult the builder as to the reliability and promptness of their work. After doing all you can think of to safeguard your investment, hire the subcontractor who best suits you.

You are now the subcontractor's boss. He is working for you, and his best performance is necessary for the completion of your project. You are the overseer and must watch all work to see that it is all right.

Be sure both of you know exactly what he is expected to do. A contract may be lacking in spelling out things that are assumed by either party. You want the subcontractor to know how much he is expected to do and how much you are going to pay him to do it.

For example, framing a house very often includes installing frame lumber, hanging windows and exterior doors, fascia board, soffit, and even gable ends. The carpenter knows this is the usual job, and the homeowner assumes it, but if it is not spelled out in a contract, the carpenter finishes the framing, doors, fascia, and soffit but fails to do the gable ends. The owner calls him back, but he refuses to return unless it is for extra pay. "The house is framed!" he says, and he's right. But the job of the framing contractor is not done. To avoid this scenario, make sure both you and the tradesman know everything that is expected of him, and how much money is expected of you when he does it.

Detail of the Dayton house. It may be cheaper to hire a contractor to install gutters and downspouts than for you to buy the materials and tools and do it yourself.

MORTGAGE PAYMENT TABLE

This mortgage payment table shows the monthly interest rates for various mortgage amounts at different terms.

20-YEAR MORTGAGE

AMOUNT	AT 7%	7½%	8%	8½%	9%
$10,000	77.60	80.60	83.60	86.80	90.00
12,000	93.12	96.72	100.32	104.16	108.00
14,000	108.64	112.84	117.04	121.52	126.00
16,000	125.16	128.96	133.76	137.68	144.00
18,000	138.68	145.08	150.48	156.24	162.00
20,000	155.20	161.12	167.20	173.60	180.00
24,000	186.24	193.44	200.64	208.32	216.00
28,000	216.28	225.68	234.08	243.04	252.00
30,000	232.80	241.80	250.80	260.40	270.00
36,000	279.36	290.16	300.96	312.48	324.00

25-YEAR MORTGAGE

AMOUNT	AT 7%	7½%	8%	8½%	9%
$10,000	70.70	73.90	77.18	80.52	83.92
12,000	84.84	88.68	92.62	96.63	100.70
14,000	98.98	103.46	108.05	112.73	117.49
16,000	112.96	118.24	123.49	128.84	134.27
18,000	127.26	133.02	138.93	144.94	151.06
20,000	141.40	147.80	154.36	161.05	167.84
24,000	169.68	177.36	185.24	193.25	201.41
28,000	197.96	206.92	216.11	225.46	234.97
30,000	212.10	221.70	231.54	241.57	251.76
36,000	254.52	266.04	277.85	289.88	302.11

30-YEAR MORTGAGE

AMOUNT	AT 7%	7½%	8%	8½%	9%
$10,000	66.60	69.92	73.38	76.89	80.46
12,000	78.92	83.91	88.05	92.27	96.55
14,000	93.24	97.89	102.73	107.65	112.65
16,000	106.56	111.87	117.40	123.03	128.74
18,000	119.88	125.86	132.08	138.40	144.83
20,000	133.20	139.84	146.75	153.78	160.92
24,000	159.84	167.81	176.10	184.54	193.11
28,000	186.48	195.78	205.45	215.30	225.29
30,000	199.80	209.76	220.13	230.67	241.39
36,000	239.76	251.72	264.16	276.81	289.66

RELEASE OF LIEN

DAYTON, OHIO, ____________________ 19____

To CITIZENS FEDERAL SAVINGS & LOAN ASSOCIATION of Dayton, Ohio

FOR VALUE RECEIVED, We hereby release all right of lien which we now have or may hereafter acquire by reason of labor or material furnished or to be furnished by us for

____

on the house and other improvements constructed for

____

on ______________________________________ Street, in Dayton, Ohio, on the following lot of land of which he is the owner, to-wit: ________________

____

____

The lending institution requires a lien release each time you pay a subcontractor during the construction of your home.

Chapter 8

Securing Your Construction Loan

The best place to obtain a loan to construct your new home is from your local savings and loan association. If one is not available, the bank in your area is your next best bet. A savings and loan where you have done business before is always a good place to start.

Selling yourself and your idea is what you have to do when you approach an institution to seek a construction loan. But knowing what will happen is only half your battle. The other half is to convince the savings and loan that you can make it happen.

You may not be able to convince the first place you try that, although you are not a carpenter or in the building business, you are very capable of building your own home.

According to one savings and loan officer who has been doing construction loans in a large metropolitan area for the past eight years, only one out of ten who attempt this project is capable of finishing it. To be that one out of ten, you must prepare yourself with all the information needed to obtain your construction loan.

When you request the forms needed to obtain the loan, the savings and loan association will give you several documents. These may be forms you have never seen before, but they will be similar to the ones illustrated in this book.

One of the forms you must complete is a personal confidential financial statement. Don't hesitate to prepare this form yourself. The form simply states what you own and what you owe and is necessary for the approval of your loan. Complete this form in your home and do one line at a time. Get a couple of copies of this document and all other documents that you need to fill out.

The personal financial statement in this book can be used as a practice guide to familiarize yourself with the information that may be needed to secure your loan. Jot down the figures that pertain to your personal financial position.

Starting with the asset side of the form, list your current bank balance in your checkbook.

Next list any savings you may have in banks, savings and loan associations, and credit unions.

List the cash value of any U.S. Savings Bonds you have. Cash value is the amount you actually paid for the savings bonds.

In "Schedule I" list any stock you have or any other securities you can sell and turn into quick cash. List any bonds you have under "Other Securities."

In "Schedule II" list the people who owe you money in the form of accounts receivable or notes. If you have loaned money to friends or relatives, list the amount in the appropriate space.

List in "Schedule III" the life insurance you carry, and if there is any cash value, put that amount on the line that says, "Cash Value Life Insurance." Whether or not your life insurance has a cash value, it is best to complete "Schedule III" just to give additional information to the loan officer.

In "Schedule IV" list the real estate you own. Don't forget to include the vacant land you plan to build your new house on.

List the make, year, and value of the automobiles you own.

Under "Other Assets," list any other things you own. You may own a boat, a motorcycle, diamonds, business equipment, snowmobile, tractor, recreational vehicles, antiques, coins, paintings,

PLEASE COMPLETE FULLY

CONFIDENTIAL FINANCIAL STATEMENT

NAME ______________________________ DATE OF STATEMENT ____________

ADDRESS ______________________________

To: CITIZENS FEDERAL SAVINGS AND LOAN ASSOCIATION, Dayton, Ohio

FOR THE PURPOSE OF PROCURING CREDIT FROM YOU FROM TIME TO TIME, I FURNISH YOU WITH THE FOLLOWING FINANCIAL STATEMENT AND INFORMATION. AS OF THE ABOVE DATE, AND I AGREE TO AND WILL NOTIFY YOU AT ONCE IN WRITING OF ANY MATERIAL UNFAVORABLE CHANGE IN THIS STATEMENT OR INFORMATION. IN THE ABSENCE OF SUCH NOTICE, OR IF A NEW STATEMENT IS NOT SUBMITTED TO YOU, THIS MAY BE CONSIDERED A SUBSTANTIALLY CORRECT CONTINUING STATEMENT. I FULLY AGREE THAT UPON APPLICATION FOR FURTHER CREDIT, THIS STATEMENT SHALL HAVE THE SAME FORCE AND EFFECT AS IF DELIVERED AS AN ORIGINAL STATEMENT OF MY FINANCIAL CONDITION AT THE TIME SUCH FURTHER CREDIT IS REQUESTED. IN CASE OF FAILURE TO NOTIFY YOU AS AGREED ABOVE, OR IF THIS STATEMENT IS FALSE IN ANY MATERIAL RESPECT, YOU MAY DECLARE DUE AND PAYABLE ALL MY OBLIGATIONS TO YOU.

ASSETS	DOLLARS		CTS.	LIABILITIES	DOLLARS		CTS.
CASH IN BANKS AND SAVINGS & LOANS (ITEMIZE)	XX	XXX	XX	LOANS FROM BANKS AND SAVINGS & LOANS	XX	XXX	XX
CHECKING ACCOUNT				SECURED			
SAVINGS ACCOUNT				UNSECURED			
				MORTGAGE LOANS ON REAL ESTATE (SCHEDULE 4)			
U. S. SAVINGS BONDS - CASH VALUE				CHATTEL MORTGAGES & OTHER LIENS OWED			
READILY MARKETABLE SECURITIES (SCHEDULE 1)				OWED TO FRIENDS AND RELATIVES			
OTHER SECURITIES (SCHEDULE 1)				LOANS ON LIFE INSURANCE			
GOOD ACCOUNTS & NOTES DUE ME (SCHEDULE 2)				OPEN ACCOUNTS UNPAID			
DUE FROM FRIENDS AND RELATIVES				INCOME TAXES OWED			
CASH VALUE LIFE INSURANCE (SCHEDULE 3)				OTHER TAXES OWED			
REAL ESTATE OWNED (SCHEDULE 4)				ANY OTHER DEBT (ITEMIZE)			
AUTOS. MAKE YEAR							
OTHER ASSETS (ITEMIZE)							
				TOTAL LIABILITIES			
				NET WORTH (TOTAL ASSETS MINUS TOTAL LIABILITIES)			
TOTAL ASSETS				TOTAL			

ANNUAL INCOME				CONTINGENT LIABILITIES			
SALARY				AS ENDORSOR, COMAKER OR GUARANTOR			
BONUS & COMMISSIONS				ON LEASES OR CONTRACTS			
DIVIDENDS				AS BONDSMAN OR SURETY FOR OTHERS			
REAL ESTATE INCOME				OTHER			
OTHER INCOME							
TOTAL INCOME							

PERSONAL INFORMATION	GENERAL INFORMATION
BUSINESS OR OCCUPATION AGE	ARE ANY ASSETS PLEDGED OR HYPOTHECATED?
	ARE YOU DEFENDANT IN ANY SUITS OR LEGAL ACTIONS?
NAME OF ANY OTHER VENTURE IN WHICH PARTNER OR OFFICER:	INDICATE ABOVE AND ON REVERSE SIDE ANY ASSETS NOT IN YOUR NAME
WIFE'S NAME SINGLE ☐	
NUMBER OF CHILDREN NUMBER OF DEPENDENTS	HAVE YOU EVER TAKEN BANKRUPTCY?

I HEREBY CERTIFY THAT THE ABOVE STATEMENT AND REPRESENTATIONS AND THE SCHEDULES AND REPRESENTATIONS ON THE OTHER SIDE OF THIS SHEET ARE TRUE AND CORRECT AS OF THE ABOVE DATE, AND I UNDERSTAND THAT YOU WILL RELY ON SUCH INFORMATION WHEN EXTENDING OR MAINTAINING CREDIT.

DATE SIGNED ____________ 19______ (SIGNATURE) ______________________

(PLEASE COMPLETE THE SCHEDULES ON THE OTHER SIDE.)

IF ADDITIONAL SPACE IS NEEDED USE SEPARATE SHEET AND SIGN

SCHEDULE 1 - SECURITIES OWNED (DO NOT INCLUDE U. S. SAVINGS BONDS)

NO. SHARES OF STOCK / PAR VALUE OF BONDS	DESCRIPTION	PFD. / COM.	CURRENT VALUE		WHERE TRADED	REGISTERED IN NAME OF
			PER UNIT	AMOUNT		

SCHEDULE 2 - GOOD ACCOUNTS AND NOTES DUE ME

MAKER	ACCT. OR NOTE	DATE DUE	ORIGINAL AMOUNT	BAL. DUE	COLLATERAL, IF ANY

HAVE ANY OF ABOVE NOTES BEEN DISCOUNTED?

SCHEDULE 3 - LIFE INSURANCE CARRIED

ISSUING COMPANY	BENEFICIARY	KIND OF INSURANCE	FACE AMOUNT	CASH VALUE	AMOUNT OF POLICY LOAN

SCHEDULE 4 - REAL ESTATE OWNED

LOCATION & DESCRIPTION	TITLE IN NAME OF	DATE ACQUIRED	COST	MARKET VALUE	MORTGAGE		
					BAL. DUE	DATE DUE	HELD BY

REMARKS:

A confidential financial statement used by lending institutions.

or other expensive art work. If you own any or all of these or other items which have value, they should all be listed at their fair market value.

The liability side of the financial statement is where you list all the debts you owe. List all loans from banks and savings and loans, whether secured or unsecured.

In "Schedule IV" list mortgage loans on real estate you own.

List any chattel mortgages and other liens. A chattel mortgage could be the amount you owe on your automobile or other assets.

List the amount of loans owed to friends and relatives.

If you have made a loan against your life insurance, list that.

List the amount of consumer loans on which you owe a balance.

If you owe any income tax, list that amount, and on the next line list any other taxes you may owe that are current and unpaid.

List other debts you may have, and itemize each one.

Under "Annual Income" list your salary for one year. Then list any bonus or commissions you will receive.

List dividends you receive from stocks you own.

If you have rental property, you should list that income under "Real Estate Income."

Under "Other Income" list earnings from a second job, spouse's income, interest income, or anything else.

In the "Contigent Liabilities" section, you need to list any notes you have co-signed as an endorser, co-maker, or guarantor.

List any leases or contracts you are responsible for. If you have signed as a bondsman or surety for others, list that. Any other liabilities you may be responsible for should be listed as well.

In the "Personal Information" section, list your business or occupation and age. Name any other venture in which you are a partner or officer. List your spouse's name if you have one. If you are not married, check the "single" block. List the number of children you have. Also list any other dependents you have, if any.

The "General Information" section should be completed next. Is any part of your assets pledged or hypothecated? This means, have you borrowed money and used your assets as collateral?

Answer "yes" or "no" if you are a defendant in any suits or legal actions.

Are all the assets you have listed in your name? If not, list the ones you may co-own. Indicate above and on the reverse side any assets not in your name.

Have you ever claimed bankruptcy? If so, give particulars.

After you have completed the Confidential Financial Statement, date the form and sign your name. This statement is the most important of the loan forms you must complete. After you have filled it out, you should then complete an estimate sheet on the cost of the home you want to build. You should have all the information you need from the estimating you did earlier on the cost of your new home. Different savings and loans use different estimate forms, but the one illustrated in this book covers all the information and can be transferred to another form very easily.

You also need to complete a mortgage loan application. This supplies information which will be needed when your short-term construction loan becomes a long-term mortgage loan after the house is completed.

The savings and loan will probably supply you with a specification sheet for new house or construction loans. Complete this form, and retain a copy for your files.

The savings and loan will also supply you with an inspection sheet. This is a check-off list for the different portions of your house as they are completed.

Construction loans are usually set up in one of two ways, either by "draw" or "direct pay." The draws are payments given to you to pay for supplies, materials, and contract help after each specific part of the work has been completed. Draws are usually delivered in four installments: one sixth, one sixth, one third, and one third of the total loan. An inspector from the savings and loan will come to the job site to see that the work you say has been completed has indeed been done.

When you request a draw from your lending institution, you may need to have affidavits and lien releases for all the work you are requesting money for. It is necessary to have a lien release signed by the subcontractor each time you pay for material, labor, or subcontract work. The savings and loan will probably give you a form similar to the one in this book, explaining the type of disbursement method you will use.

You may be required to fill out an affidavit and sign it each time you request a draw. This affidavit states who you are, what work has been completed, who did the work, what materials were used, what the costs were, how much you have paid, and the amount that is still owed to the subcontractor.

The affidavit in this book is used by some institutions or material suppliers, but there are other affidavits which are similar.

If you can obtain a direct pay plan for your construction loan, you should request this type. In a direct pay plan you sign the invoice for material, supplies, or subcontracting work and mail it to the loan institution. The lender will inspect your job site to verify the invoice, will then pay the invoice, and get a release or lien from your vendor. This way you do not have to write any checks or get any lien releases.

Most building material suppliers prefer the direct pay system as they are more sure of collecting their money from a savings and loan well known in the area, compared to someone who is building only one house. Building suppliers can sometimes mail the invoices directly to the lending institution and again save you an extra step.

The fee for this direct payment is minimal, usually between $25 and $50. Can you afford this little expense for a big job?

Most construction loans are for six months. At the end of that time, if your new home is not completed, you can request an extension of time to complete your project. An extension will be granted if you have completed a reasonable portion of your building and show good faith that you will finish the rest in a reasonable amount of time. According to the time needed, your savings and loan will probably grant you up to an additional ninety days. This should give you enough time to complete your project.

You may be requested to put down a cash payment on your home loan even though your land is paid for. Any construction costs you pay count toward your down payment. If you can get a direct pay loan you will be required to deposit your down payment with the savings and loan, which will be used first in paying for construction. This way you don't start using the borrowed money until your own cash has been spent, and the amount of interest you have to pay is decreased.

The typical construction loan of $40,000, with a land value of $9,000, may require a down payment of $800 to $4,000, depending on the lending institution. Remember, the more you can get together to pay down, the lower your interest payments will be.

If you don't have a direct pay system and have to pay the invoice yourself, open a special checking account. This account should be used only for your house project and will enable you to keep more accurate and complete records.

At the completion of your building project, your construction loan will, in most cases, convert to a mortgage loan because you have supplied all the information needed at the time you applied for your construction loan. You were probably asked to sign an open-end mortgage which has variances for the decrease or increase of the principal amount, according to the needs of your new home construction. This is an advantage to you. If you should go over your estimate, the savings and loan will make more money available to you, but if you should not need as much money as you first determined, you have to take only the amount you need. This saves you interest.

Once you have come to this point you are ready to begin the actual physical construction. Planning is the most important part of saving money in building your own home, but it is also important to make your plan work.

You must keep daily and weekly summaries of your cost. Knowing what your new home will cost, with changing prices, is of great importance. Take a calendar and plot off the time allowed for each phase of your building. Time is money—your money—and any delay will cost you additional money. You will be making monthly interest payments on the money you have used on your project, so keep your job moving.

RESIDENTIAL LOAN APPLICATION FOR CITIZENS FEDERAL SAVINGS & LOAN ASSN. OF DAYTON

MORTGAGE APPLIED FOR	☐ Conventional ☐ FHA ☐ VA ☐ ______	Amount $	Interest Rate %	No. of Months	Monthly Payment Principal & Interest $	Escrow/Impounds (to be collected monthly) ☐ Taxes ☐ Hazard Ins. ☐ MI ☐ ______

Prepayment Option

SUBJECT PROPERTY

Property Street Address | City | County | State | Zip | No. Units

Legal Description (Attach description if necessary) | Year Built

Purpose of Loan: ☐ Purchase ☐ Construction-Permanent ☐ Construction ☐ Refinance ☐ Other (Explain)

Complete this line if Construction-Permanent or Construction Loan | Lot Value Data | Year Acquired ______ | Original Cost $ | Present Value (a) $ | Cost of Imps. (b) $ | Total (a+b) $ | ENTER TOTAL AS PURCHASE PRICE IN DETAILS OF PURCHASE

Complete this line if a Refinance Loan | Year Acquired | Original Cost $ | Amt. Existing Liens $ | Purpose of Refinance | Describe Improvements [ ] made [ ] to be made | Cost: $ ______

Title Will Be Held In What Name(s) | Manner In Which Title Will Be Held

Source of Down Payment and Settlement Charges

This application is designed to be completed by the borrower(s) with the lender's assistance. The Co-Borrower Section and all other Co-Borrower questions must be completed and the appropriate box(es) checked if ☐ another person will be jointly obligated with the Borrower on the loan, or ☐ the Borrower is relying on income from alimony, child support or separate maintenance or on the income or assets of another person as a basis for repayment of the loan, or ☐ the Borrower is married and resides, or the property is located, in a community property state.

BORROWER

Name | Age | School Yrs ___

Present Address No. Years ______ ☐ Own ☐ Rent

Street ______

City/State/Zip ______

Former address if less than 2 years at present address

Street ______

City/State/Zip ______

Years at former address ______ ☐ Own ☐ Rent

Marital Status ☐ Married ☐ Separated ☐ Unmarried (incl. single, divorced, widowed) | Dependents other than listed by Co-Borrower: No. | Ages

Name and Address of Employer | Years employed in this line of work or profession? ______ years | Years on this job ______ | ☐ Self Employed*

Position/Title | Type of Business

Social Security Number *** | Home Phone | Business Phone

CO-BORROWER

Name | Age | School Yrs ___

Present Address No. Years ______ ☐ Own ☐ Rent

Street ______

City/State/Zip ______

Former address if less than 2 years at present address

Street ______

City/State/Zip ______

Years at former address ______ ☐ Own ☐ Rent

Marital Status ☐ Married ☐ Separated ☐ Unmarried (incl. single, divorced, widowed) | Dependents other than listed by Co-Borrower: No. | Ages

Name and Address of Employer | Years employed in this line of work or profession? ______ years | Years on this job ______ | ☐ Self Employed*

Position/Title | Type of Business

Social Security Number *** | Home Phone | Business Phone

GROSS MONTHLY INCOME				MONTHLY HOUSING EXPENSE **			DETAILS OF PURCHASE	
Item	Borrower	Co-Borrower	Total	Rent	$ Present	Proposed		
Base Empl. Income	$	$	$	First Mortgage (P&I)		$	a. Purchase Price	$
Overtime				Other Financing (P&I)			b. Total Closing Costs (Est.)	
Bonuses				Hazard Insurance			c. Prepaid Escrows (Est.)	
Commissions				Real Estate Taxes			d. Total (a + b + c)	$
Dividends/Interest				Mortgage Insurance			e. Amt. This Mortgage	( )
Net Rental Income				Homeowner Assn. Dues			f. Other Financing	( )
Other† (Before completing, see notice under Describe Other Income below.)				Other			g. Present Equity in Lot	( )
				Total Monthly Pmt.	$	$	h. Amount of Cash Deposit	( )
				Utilities			i. Closing costs paid by Seller	( )
Total	$	$	$	Total	$	$	j. Cash Reqd. For Closing (Est.)	$

DESCRIBE OTHER INCOME

B—Borrower C—Co-Borrower	NOTICE:† Alimony, child support, or separate maintenance income need not be revealed if the Borrower or Co-Borrower does not choose to have it considered as a basis for repaying this loan.	Monthly Amount
		$

IF EMPLOYED IN CURRENT POSITION FOR LESS THAN TWO YEARS COMPLETE THE FOLLOWING

B/C	Previous Employer/School	City/State	Type of Business	Position/Title	Dates From/To	Monthly Income
						$

THESE QUESTIONS APPLY TO BOTH BORROWER AND CO-BORROWER

If a "yes" answer is given to a question in this column, explain on an attached sheet.	Borrower Yes or No	Co-Borrower Yes or No
Have you any outstanding judgments? In the last 14 years, have you been declared bankrupt?		
Have you had property foreclosed upon or given title or deed in lieu thereof?		
Are you a co-maker or endorser on a note?		
Are you a party in a law suit?		
Are you obligated to pay alimony, child support, or separate maintenance?		
Is any part of the down payment borrowed?		

	Borrower Yes or No	Co-Borrower Yes or No
Do you have health and accident insurance?		
Do you have major medical coverage?		
Do you intend to occupy this property?		
Will this property be your primary residence?		
Have you previously owned a home?		
Sales Price of previously owned home?	$	$

*FHLMC requires self employed to furnish signed copies of one or more most recent Federal Tax Returns or audited Profit and Loss Statements. FNMA requires business credit report, signed Federal Income Tax returns for last two years, and, if available, audited P/L plus balance sheet for same period.

**All Present Monthly Housing Expenses of Borrower and Co-Borrower should be listed on a combined basis.

***Neither FHLMC nor FNMA requires this information.

FHLMC 65 Rev. 3/77 OLSA–FNMA 1003 Rev. 3/77

This Statement and any applicable supporting schedules may be completed jointly by both married and unmarried co-borrowers if their assets and liabilities are sufficiently joined so that the Statement can be meaningfully and fairly presented on a combined basis; otherwise separate Statements and Schedules are required (FHLMC 65 A/FNMA 1003A). If the co-borrower section was completed about a spouse, this statement and supporting schedules must be completed about that spouse also.

☐ Completed Jointly ☐ Not Completed Jointly

STATEMENT OF ASSETS AND LIABILITIES

ASSETS

Description	Cash or Market Value
Cash Deposit Toward Purchase Held By	$
Checking and Savings Accounts (Show Names of Institutions/Acct. Nos.)	
Stocks and Bonds (No./Description)	
Life Insurance Net Cash Value *Face Amount ($)*	
SUBTOTAL LIQUID ASSETS	$
Real Estate Owned *(Enter Market Value from Schedule of Real Estate Owned)*	
Vested Interest in Retirement Fund	
Net Worth of Business Owned (ATTACH FINANCIAL STATEMENT)	
Automobiles (Make and Year)	
Furniture and Personal Property	
Other Assets (Itemize)	
TOTAL ASSETS	A. $

LIABILITIES AND PLEDGED ASSETS

Indicate by (*) those liabilities or pledged assets which will be satisfied upon sale of real estate owned or upon refinancing of subject property

Creditors' Name, Address and Account Number	Acct. Name if Not Borrower's	Mo. Pmt. and Mos. left to pay	Unpaid Balance
Installment Debts (including "revolving" charge accts.)		$ Pmt./Mos. /	$
		/	
		/	
		/	
		/	
		/	
Automobile Loans		/	
Real Estate Loans		X	
Other Debts Including Stock Pledges		/	
Alimony, Child Support and Separate Maintenance Payments Owed To		/	X
TOTAL MONTHLY PAYMENTS		$	X
NET WORTH (A minus B) $		TOTAL LIABILITIES	B. $

SCHEDULE OF REAL ESTATE OWNED (If Additional Properties Owned Attach Separate Schedule)

Address of Property (Indicate S if Sold, PS if Pending Sale or R if Rental being held for income)	⇩	Type of Property	Present Market Value	Amount of Mortgages & Liens	Gross Rental Income	Mortgage Payments	Taxes, Ins. Maintenance and Misc.	Net Rental Income
			$	$	$	$	$	$
		TOTALS →	$	$	$	$	$	$

LIST PREVIOUS CREDIT REFERENCES

⇩ B—Borrower C—Co-Borrower	Creditor's Name and Address	Account Number	Purpose	Highest Balance	Date Paid
				$	

List any additional names under which credit has previously been received ______________________

AGREEMENT: The undersigned applies for the loan indicated in this application to be secured by a first mortgage or deed of trust on the property described herein, and represents that the property will not be used for any illegal or restricted purpose, and that all statements made in this application are true and are made for the purpose of obtaining the loan. Verification may be obtained from any source named in this application. The original or a copy of this application will be retained by the lender, even if the loan is not granted.

THE APPLICANT AGREES TO MAKE MONTHLY PAYMENTS BY THE ASSOCIATION'S PREAUTHORIZED PAYMENT SYSTEM AS A CONDITION OF THE CLOSING AND DISBURSING OF THE LOAN.

EQUAL CREDIT OPPORTUNITY NOTICE

You are hereby provided the following "Equal Credit Opportunity" notice as required by Section 4112.021 of the Ohio Revised Code. "The Ohio laws against discrimination require that all creditors make credit equally available to all credit worthy customers, and that credit reporting agencies maintain separate credit histories on each individual upon request. The Ohio civil rights commission administers compliance with this law."

I/we fully understand that it is a federal crime punishable by fine or imprisonment, or both, to knowingly make any false statements concerning any of the above facts as applicable under the provisions of Title 18, United States Code, Section 1014.

______________________ Date ________ ______________________ Date ________

Borrower's Signature Co-Borrower's Signature

VOLUNTARY INFORMATION FOR GOVERNMENT MONITORING PURPOSES

If this loan is for purchase or construction of a home, the following information is requested by the Federal Government to monitor this lender's compliance with Equal Credit Opportunity and Fair Housing Laws. The law provides that a lender may neither discriminate on the basis of this information nor on whether or not it is furnished. Furnishing this information is optional. If you do not wish to furnish the following information, please initial below.

BORROWER: I do not wish to furnish this information (initials) ________

RACE/NATIONAL ORIGIN ☐ American Indian, Alaskan Native ☐ Asian, Pacific Islander ☐ Black ☐ Hispanic ☐ White ☐ Other (specify) ________ **SEX** ☐ Female ☐ Male

CO-BORROWER: I do not wish to furnish this information (initials) ________

RACE/NATIONAL ORIGIN ☐ American Indian, Alaskan Native ☐ Asian, Pacific Islander ☐ Black ☐ Hispanic ☐ White ☐ Other (specify) ________ **SEX** ☐ Female ☐ Male

FOR LENDER'S USE ONLY

(FNMA REQUIREMENT ONLY) This application was taken by ______________________ (Interviewer), a full time employee of **Citizens Federal Savings & Loan of Dayton**, (Name of Lender) in a face to face interview with the prospective borrower.

Date ____ Amt. $ ______ Int. Rate ____% Yrs. ____ Mo. Pay. ______ Escrow Tax ______ Haz. Ins. ______ PMI. ______

Blanket Loan - Amt. $ ______ Int. Rate ____% Yrs. ____ Verify Emp. ______ Verify Funds ______

Commitment Expires ______ % to Appraisal ______ % to Purchase Price ______ Commitment Fee $ ______

Members Approving ______________________

Remarks ______________________ Income Ratio P & I (___) Yrs.

Estimated Tax & Ins. ______

Estimated Total Payment ______

Ratio ______ To 1 ______

Amt. Monthly Income ______

Over & Under ______

FHLMC 65 Rev. 3/77 REVERSE OLSA–FNMA 1003 Rev. 3/77

An application like this one is necessary when applying for a construction loan.

LABOR

Said affiant further says that the following shows the names of every unpaid laborer in the employ of (2) ____________________ furnishing labor under said contract, giving the amount, if any, which is due, or to become due, for labor done to date hereof.

NOTE—If the fact is that every laborer has been paid in full, then recite: "Every laborer has been paid in full." If not, then give each unpaid laborer's name and amount due or to become due.

NAME	HOURS	Amount Due or to Become Due for Labor Furnished to Date Hereof	

That the amounts due or to become due to said sub-contractors, material-men and laborers, for work done or machinery, material or fuel furnished to the date hereof, to (2) ____________________ are fully and correctly set forth opposite their names, respectively, in the aforesaid statements, and further evidenced by certificates of every person furnishing machinery, material or fuel, hereto attached, and made a part hereof.

Affiant further says that ____________________ has not employed or purchased or procured machinery, material or fuel from, or sub-contracted with any person, firm or corporation, other than those above mentioned, and owes for no labor performed, or machinery, material or fuel furnished, under said contracts, other than above set forth.

____

SWORN TO BEFORE ME AND SUBSCRIBED IN MY PRESENCE, at ____________________ Ohio,

this __________ day of ____________________, A. D. 19______.

____

NOTARY PUBLIC, __________ COUNTY, OHIO.

1. Sec'y., Treas., one of firm, or agent, as case may be.
2. Name.
3. "Owner," "part-owner," "lessee," or "authorized agent of the owner, part-owner or lessee," or "original" or "principal contractor under a contract with .... the owner, part-owner or lessee." as the case may be
4. "Constructing, altering or repairing a boat, vessel or other watercraft," or "erecting, altering, repairing or removing a house, mill, manufactory, or any furnace, or furnace material therein, or other building appurtenance, fixture, bridge or other structure," or "digging, drilling, boring, operating, completeing and repairing of any gas well, oil well or other well," or "altering, repairing or constructing any oil derrick, oil tank, oil or gas pipe line," or "furnishing tile for the drainage of any lot or land," as the case may be.
5. Accurate description of property.

AFFIDAVIT

of

ORIGINAL OR SUB-CONTRACTOR

____________________ OWNER

____________________ SUB-CONTRACTOR

____________________ HEAD-CONTRACTOR

DATE ____________________ 19______

An affidavit signed by the homeowner at the time he seeks a draw from the lender.

Affidavit of Original or Sub-Contractor.

The State of Ohio, ______ *County,* } *ss.* ______, *Ohio,* ______ *19*___

______, *being first duly sworn, says that he is (1)* ______

of (2) ______ *the* { Sub / Original } *contractor having a contract with (2)*

______ *the (3)* ______

______ *for (4)* ______

a ______ *situated on or around or in front of the following described property (5)*

in ______ *County, Ohio, viz:* ______

______

______

whereof (2) ______ *was the owner, part owner or lessee,*

SUB-CONTRACTORS

Affiant further says that they following shows the names of every sub-contractor in the employ of said

(2) ______ *giving the amount, if any, which is due, or to become due, to them, or any of them, for work done, or machinery, material or fuel furnished to date hereof, under said contracts.*

NOTE—This statement must be accompanied by similar sworn statement signed by each of the sub-contractors listed below.

NAME	TRADE	Amount Due or to Become Due for Work and Materials to Date Hereof.	

MATERIAL MEN

Said affiant further says that the following shows the names of every person furnishing machinery, material or fuel to (2) ______ *giving the amount, if any, which is due, or to become due, to them, or any of them, for machinery, material or fuel furnished to date hereof, under said contracts.*

NAME	Kind of Machinery, Material or Fuel	Amount Due or to Become Due for Material Furnished to Date Hereof.	

NOTE—The above must be accompanied by "CERTIFICATE OF MATERIALMAN". In lieu of such certificate, there may be furnished a written waiver of lien, a written release or receipt.

CUSTOMER NAME ______________________ PHONE ____________

PROPERTY ADD. ______________________________

LOT NO. ______________________________

BUILDER ______________________ PHONE ____________

MAILING ADDRESS:

LOAN NO. ______________

CONTRACT PRICE OR
EST. CONST. COST $ ____________

LOAN AMT. $ ____________

RATIO OF LOAN/VALUE ________ %

DEPOSIT AMT. BY CUST. $ ________

TOTAL CONST. FUNDS IN L.I.P.$ ______

DISBURSEMENT RECORD:

AMT.	APPROVAL	DATE	CK. #
$			
$			
$			
$			
$			
$			
$			
$			

DISBURSEMENT METHOD:

☐ To Builder (4 draw basis)

☐ Direct Pay of Bills (4 draw basis)

☐ AIA Progress Form G702 Monthly Payout Basis.

AFFIDAVITS AND
LIEN RELEASES NEEDED:

First Draw (Affidavit Plus Releases)

___ Insurance

___ Well Report

___ Staking, Layout, Clearing Site

___ Excavate Footers, Slab, Bsmt.

___ Pour Footers

___ Foundation ___ Walls ___ Labor ___ Material

___ Pour Slab ___ Labor ___ Mat.

___ Build Deck ___ Labor ___ Mat.

Second Draw (Affidavit Plus Releases)

___ Truss System ___ Framing Lumber

___ Carpenter Labor ___ Erection

___ Roofing ___ Labor ___ Mat.

___ Plumbing 2nd Rough ___ Labor ___ Mat.

___ Electric Rough ___ Labor ___ Mat.

___ Heat Rough ___ Labor ___ Mat.

___ Windows

Third Draw (Affidavit Plus Releases)

___ Brick Siding ___ Mat. ___ Labor

___ Drywall ___ Mat. ___ Labor

___ Insulation ___ Mat. ___ Labor

Fourth Draw (Affidavit Plus Releases)

___ Plumbing Final ___ Mat.

___ Heating Final ___ Mat.

___ Electric Final ___ Mat.

___ Cabinets ___ Vanities

___ Lumber & Mill Work

___ Carpenter Labor

___ Septic

___ Painting

___ Carpet

___ Drive & Walks

___ Grade

___ Sod or Seed

___ Shrubs

This is a record showing the progress on your house as each draw is paid. The lending institution keeps it in its files, but you should make and keep a copy, too.

COMMENTS

% Complete	Inspector	Date

____Foundation Measurements
____Excavating
____Footings
____Foundation
____Waterproofing
____Backfill
____Deck or Slab
____Framing
____Partitions
____Sheathing
____Sash
____Roof ____Tar Paper____ Shingles
____Rough Plumbing
____Rough Wiring
____Rough Heat
____Water or Well
____Sewer or Septic
____Fireplace
____Exterior Walls Brick ____Siding
____Insulation
____Lath Hung
____Plaster or Spackel
____Basement Floor Poured
____Trim
____Doors (Interior)
____Kitchen Cabinets
____Tile Work Baths____Kitchen
____Final Plumbing
____Final Wiring
____Final Heating
____Carpet
____Painting (Exterior)
____Painting (Interior)
____Stoops
____Walks - Patio
____Driveway
____Final Grade
____Garage Completed
____Sod____ Seeded
____Landscaping ____Final Grade____Shrubs ____Trees

x Fully Complete ✓ - Partially Complete

(Photo) **1st Draw**	**(Photo)** **2nd Draw**
(Photo) **3rd Draw**	**(Photo)** **Final Draw**

ZONING CERTIFICATE
COUNTY OF MONTGOMERY
Division of Building & Zoning Inspection

TO: ____
NAME

Address ____

Date ______________________ 19____ No. 6379

THIS IS TO CERTIFY, That the ____

located at the ________ side of ______________________________ and ________ ft. ________ from

intersection of ______________________________ lot ________ subdivision ______________________

township ______________________ section ________ zone ________ is approved for use as ______________________

______________________________. This is a ______ conforming, ______ non-conforming*, ______ special*, use under

Article ______ Section ________ of the Montgomery County Zoning Resolution. *See Appeal Board Case No. ______________.

Restrictions or remarks: ____

A zoning certificate is usually required even before the building permit and whether or not your home is in compliance with local statutes.

This is an application for a plumbing permit. The "B-1-2-3-4" columns refer to the floor of the building. The fees may vary for this permit, depending on how many fixtures are to be installed, kind of fixtures, and the number of drains.

MONTGOMERY COUNTY
HEALTH DISTRICT
DIVISION OF PLUMBING
451 West Third Street Dayton, Ohio 45402

APPLICATION FOR A PLUMBING PERMIT

PERMIT NO. ______________________
FEE ______________________
DATE ______________________ 19____

APPLICATION IS HEREBY MADE FOR A PERMIT TO DO PLUMBING AS DESCRIBED IN THIS APPLICATION AND TO BE INSTALLED IN ACCORDANCE WITH THE PLUMBING CODE OF MONTGOMERY COUNTY.

LOCATION ______________________ MUNICIPALITY OR TOWNSHIP ______________________ BUILDING NEW ☐ OLD ☐

OWNER ______________________ INSTALLATION NEW ☐ EXTENSION ☐ REPLACEMENT ☐

OWNER'S ADDRESS ______________________

BUILDING USED AS ______________________ HOUSE DRAIN EMPTIES ______________________

FIXTURES	B	1	2	3	4	FIXTURES	B	1	2	3	4	FIXTURES	B	1	2	3	4	FIXTURES			
WATER CLOSETS						AUTOMATIC HEATER						DRINKING FOUNTAIN									
LAVATORIES						SHOWER BATH						SODA FOUNTAINS									
BATH TUBS						URINALS						STEAM TABLES									
SINKS						FLOOR DRAINS						SLOP SINKS									
WASH TRAYS						SOFTENERS						DENTIST CUSPIDORS									
GARBAGE DISPOSALS						SUMPS						FROST PROOF CLOSETS									
DISH WASHERS						GARAGE DRAINS						BAR WASTE									
AUTOMATIC WASHERS						DOWN SPOUTS						SEPTIC TANK						TOTAL FIXTURES			

DO NOT WRITE IN THIS SPACE

	DATE	INSPECTOR
ROUGH		
FINAL		

APPLICANT ______________________

ADDRESS ______________________ PHONE ____________

APPROVED BY ______________________ ISSUED BY ____________

Chapter 9

Buying Your Permits

Buying all the permits you need can be a problem if you don't know the procedure you must follow. Consult the building regulations department in your area to determine what permits will be necessary and where you should begin. Procedure varies in different counties and in different states, but if you are building in or near a large metropolitan area, your procedure will be similar to what follows.

The first step is usually getting approval from the environmental health department which has jurisdiction over anything you do 3 to 5 feet or more from your house. This department controls approval of your septic tank, well, cistern, swimming pool, and any other permanent construction you may do on your property. The people who work for this department can be helpful or harmful if political appointees let whim rule their judgment.

In Greene County, Ohio, for example, the environmental health department is uncooperative and can stop your building before you begin if you don't adhere strictly to its interpretation of archaic rules and regulations.

The Greene County Environmental Health Department resents the outflow of population from Dayton, a large metropolitan area, into Greene County's farming communities. Therefore, if you are moving out of a large metropolitan area into the suburbs, across county lines, be prepared for an unfriendly reaction, especially if the flow into the area is new. Don't let this discourage you. Someone has to be first.

Get your permit for a septic tank, and then you are ready for your well and pump permit. Once you have these, your next stop is your zoning board.

What does your zoning board do? Your zoning board will approve your plot plan and sell you a zoning permit. It will also advise you if your plot plan is wrong. Most zoning boards like to keep houses in a straight line. That is, they are the ones who determine how far from the street your house should be on your lot. They also determine how near the side edge of your property line you may build your house and where you can build your driveway, as well as selling you a permit for it.

Once you have completed your business with the zoning board, you can go to the building regulations department to buy your building permit.

The building regulations department usually requires you to submit one of your two sets of building plans for them to keep. This set will be examined by an inspector to verify that your plans meet with building regulations. If there are any exceptions, they will notify you by mail and ask you to come into their office to discuss the corrections needed, but this is very unusual. Any adjustments that are needed in your plans are noted in red and returned to you. The set of plans approved by the building department must be kept on your job site. If you have any questions concerning the notes made by the building regulations department, you should call and discuss any changes with them.

Once you have gotten your building permit, you are then allowed to buy your electrical permit. This permit is purchased from the electrical department.

The next permit to be obtained is your heating and air conditioning permit, usually purchased from the building regulations department.

Your plumbing permit is next.

Remember that each time you buy a permit,

County of Montgomery
DIVISION OF BUILDING INSPECTION
451 W. 3rd STREET, DAYTON, OHIO 45402

PERMIT

Void Unless Stamped "PAID" by Division

YELLOW CARD MUST BE POSTED ON PREMISES

Void after six Months from Date of Issue

PERMISSION IS HEREBY GRANTED TO — APPLICATION No.

Mr.

Address, — Phone

Location — Side of — Feet

From Inter. of — Twp. — Sec.

House No. — Lot No. — Plat

Book No. — Page — Zone — Zoning Approval

To — Story — Family. — Other

CONSTRUCT — — Fee

Type — Sq. Ft. — Valuation $ — $

Miscellaneous — Fee

Type — Sq. Ft. — Valuation $ — $

Heating — Fee

B.T.U. Rating

For Const. — Old House — Valuation $ — $

Final Inspections Required for Permanent Electric Meter

Check Your Deed and Plat Restrictions — Total Fee $

Const. Cont. — Address — Phone

Plans By — Address — Phone

Heat. Cont. — Address — Phone

All work to be done in accordance with the laws of the State of Ohio, Uniform Building Code and zoning of the County of Montgomery, the application and the plans and/or specifications as approved by the County Building and Zoning Inspector.

Owner

Address

Approved By — Issued By — As

Additional permits are required for plumbing, electrical, water and sanitation.

Date

PERMIT **No.**

Howard A. Valignette
County Building Inspector

Form No. BI 5—Revised — APPLICANT

Form No. BI 5—Revised — OFFICE COPY

Form No. BI 5—Revised — AUDITOR'S COPY

Form No. BI 5—Revised — ELECTRICAL INSP. DEPT.

Form No. BI 5—Revised — SEWER AND WATER DEPT.

Form No. BI 5—Revised — HEATING INSP. DEPT.

Form No. BI 5—Revised — BOARD OF HEALTH

BUILDING REGULATIONS DEPARTMENT
451 West Third Street
Electrical Safety Inspection
Dayton, Ohio 45422
225-4636
225-4630

READY______

WILL CALL______

ELECTRICAL PERMIT

BUILDING PERMIT No. ______

Owner______ Location of Job______

Address______ Phone______

Applicant______

Address______ Phone______

Installation: ___New ___Extension ___Industrial

Building: ___Old ___New

Bldg. Type ___Res. ___Multi-Fam. ___Comm. ___Indust.

Twp.______ Bldr.______

Territory:______ Approval to Utility:______

______ 19___
Signature of Applicant Date

______ 19___
Approved: Chief Electrical Safety Inspector Date

Void after 6 mo. unless work is started. Plans and calculations must be submitted and signed by applicant in duplicate and become a part of this application.
Permit issued to do electrical work as described herein to be installed in accordance with the National Electrical Code, Ohio Building Code, Uniform Building Code and must be signed by the Electrical Safety Inspector before starting work.

ITEM	NO.	PRICE	AMOUNT
Service (min.)		$10.00	$
Temp. Service		10.00	
Sub. Service		1.00	
Cable Heat		10.00	
2-Wire Circuit		1.00	
3-Wire Circuit		1.00	
Final Insp.		10.00	
Comm. Insp.		10.00	
Re-Inspection		10.00	
Fur./Air Cond. Unit		10.00	
Additional Units		1.00	
Outlets 1 - 10		10.00	
Outlets over 10 by circuits & Final			
Sign (only)		10.00	
Temporary Pole		10.00	
Pool Bonding Each		10.00	
Commercial Reconnect		10.00	
Hazardous Wiring		10.00	
TOTAL $			

(Over for Electrical Insp. Report)

ELECTRICAL SAFETY INSPECTION REPORT

TYPE OF INSP.	DATE	REPORT	INSPECTOR

This two-sided electrical permit application advises the particular department that you are constructing a home. On the reverse (bottom) it includes a record of inspectors.

(Left) You will become very well known in your county courthouse. Here is a seven-part carbon set permit application showing where each copy goes.

you must submit a set of your building plans for approval. The cost of your permit usually depends on your building plans.

The best way to buy your permits is to buy them all in one day, if possible. This eliminates extra trips to the courthouse or county services building.

Acquiring your own permits, rather than letting a contractor or tradesman get them for you, is one of the first and best ways for you to save money on your home. A plumber who charges $15 or $20 or more an hour to install your pipes will charge the same amount for the hours he spends driving to the courthouse and sitting while clerks and inspectors do the paperwork. By getting your own permits, you save money, the permits are just as valid, and you will find most of the office people very helpful.

Permit No.__________

DIVISION OF BUILDING INSPECTION

MONTGOMERY COUNTY, OHIO

DATE__________

LOCATION

FOR

NOTICE

This card must be left on the premises and posted close to or on the front of the building until completion.

____

County Building Inspector

HEATING

Permit No.__________

DIVISION OF BUILDING INSPECTION

MONTGOMERY COUNTY, OHIO

DATE__________

LOCATION_____SIDE_____

_____, _____FEET_____FROM INT.

____

LOT NO.__________

HEATING CONTRACTOR

____

PLEASE PRINT

FOR INSPECTION GIVE PERMIT NO.

THIS CARD MUST BE POSTED NEAR FURNACE.

Howard A. Valiquette

COUNTY BUILDING INSPECTOR

A building permit is not a heating permit.
You must get both, as well as permits to drill
a well, install plumbing, install electrical wiring,
build a pool, put in a septic tank.
Know your building code,
and be sure you have every permit required.

AFFIDAVIT

State of Ohio
County of Montgomery
Division of Plumbing
Health Department

I ____ of

____, Ohio Being

first duly sworn, depose and say that I am the owner of the single dwelling located at or being constructed at ______________________________

____, Ohio

and I am occupying or will immediately occupy the same and that all the plumbing work proposed to be installed at this location will be done by me personally according to all applicable standards and that I will call for inspections at the required stages of construction.

Upon signing this affidavit, I further affirm that I understand that the home owner's permit authorized by law and issued as the result of this application is valid for only six (6) months from the date of issuance and that only two (2) ninety (90) day extensions may be granted and these extensions must be requested in writing to this office.

I further affirm that I understand that any permit shall become invalid if the authorized work remains uninitiated, or is suspended, postponed, or abandoned throughout any six (6) month period, and that any plumbing work done other than by myself must be by a contracting plumber registered and bonded with the Montgomery County Health District and under a separate permit, and further affiant saith not.

____
Home Owner

Subscribed and sworn before me this _____ day of ____________________

____
Notary Public

Many local governments permit an unskilled plumber to work on a home provided he is the person who will live in it. If a homeowner hires a plumber, the plumber must be licensed, or the inspectors may reject all the work he does. This affidavit is filed by a homeowner who expects to work on his own home and is signed at the time he applies for a permit. There is usually a small filing fee, about 50 cents, for this document. The plumbing work must meet all code requirements and be approved by an inspector.

MAY SMOLDER

Chapter 10

Where Does the House Go?

Where does the house go? You may have very little, if anything, to say in determining where on the lot your building is located. The zoning board determines how far from the street your house should be and how near your property line you are allowed to build. Within these limits, you can determine where you want your house.

The environmental health department determines where your septic tank and well will be located. If there is no environmental health department, you can put your septic tank in the front yard and your well in the back yard, according to your preference.

Keep your well at least 15 feet from your house so the treatment of the foundation for termites won't affect your water supply. There should be an acceptable distance between the well and the septic tank to prevent seepage from your septic tank into your water system. Most areas have regulations stating how far apart the two systems should be.

It is best to notify the electric company of your intentions to build since they may need as much as three weeks or more to bring service to your house. Be sure to supply them with a plot plan to enable them to provide you with the best service. Utility companies are usually very prompt in servicing new homes, but due to the volume of new home construction in some areas, it is best to notify them in advance for this service.

The telephone company should also be notified and advised as to when you plan to have the house under roof. This enables them to schedule your house to be prewired for telephones and could save you money on later installation.

This is a vent pipe installed in the sewer or septic tank line. It is required by most health departments to prevent buildup of gasses.

Staking Out the Lot

Why do you have those sticks and strings? You need those sticks and strings to make sure the walls of your house are straight and the corners are square. This is a job that should not be done by any Tom, Dick, or Harry walking down the street. The staking-out of your house should be done by an engineer or someone who knows how to use a transit and is experienced in laying out buildings. You need the stakes in the ground as a reference point to know where you are going.

While the man is there with the transit he should put stakes at each corner of the building with a level mark on them, so you can keep your foundation level. This will also help you later to determine if your floors are level. Obviously, all of these stakes need to be protected and not moved or bent. Don't let the kids play on them.

You should also have the engineer stake out your driveway at this time, if you don't already have a driveway.

After the house has been staked out on the building lot, it is a good idea to determine where your well will be located. Check with your local health authority first, and be sure your permit is in order. Locate your well at least 15 feet from the foundation walls. Then call the welldigger and start digging your well. Drilling the well should take only a few days, but it should be done at this time to make water available during the construction of your house.

If you are in an area where the water supply is questionable, you could even drill your well before staking out the house. Some people hope to put wells in their front yards where the well head can be hidden by attractive flowers or shrubs, but the water may not be there, and your well could wind up in the back yard.

(Right)
If a piece of farmland has been placed on the market for residential use, be sure the property is surveyed by an independent surveyor or engineer. These white stakes mark the border of two, deep, five-acre plots. Stakes are also placed at each corner of the building with a level mark to help keep the foundation level.

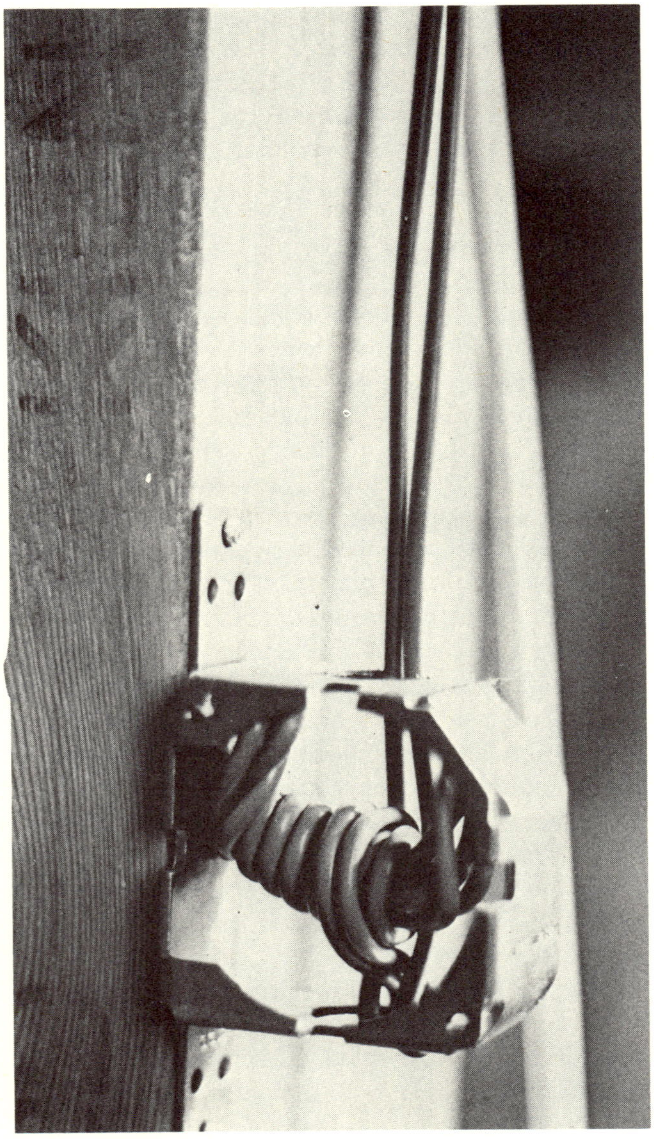

(Left)
This is a rough-wired telephone installation. The telephone company usually prewires a home under construction without charge, saving work in future years when phones are moved around. After installation, telephone jacks are very unobtrusive, so go ahead and get them in every room while it's easy to do.

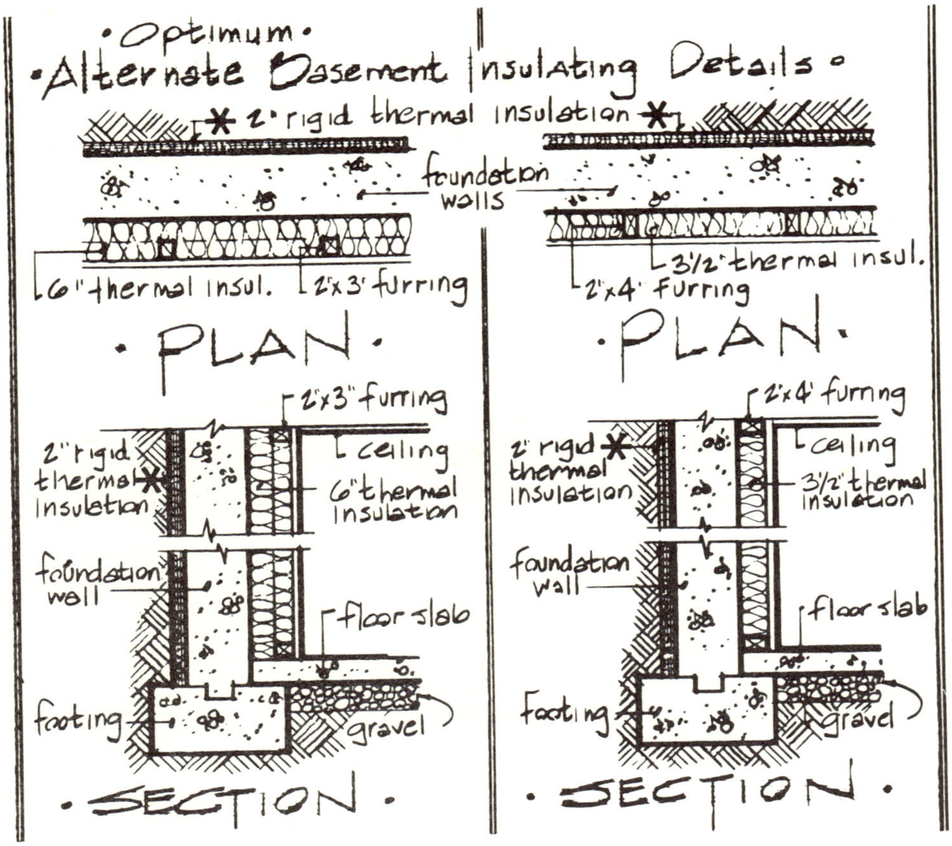

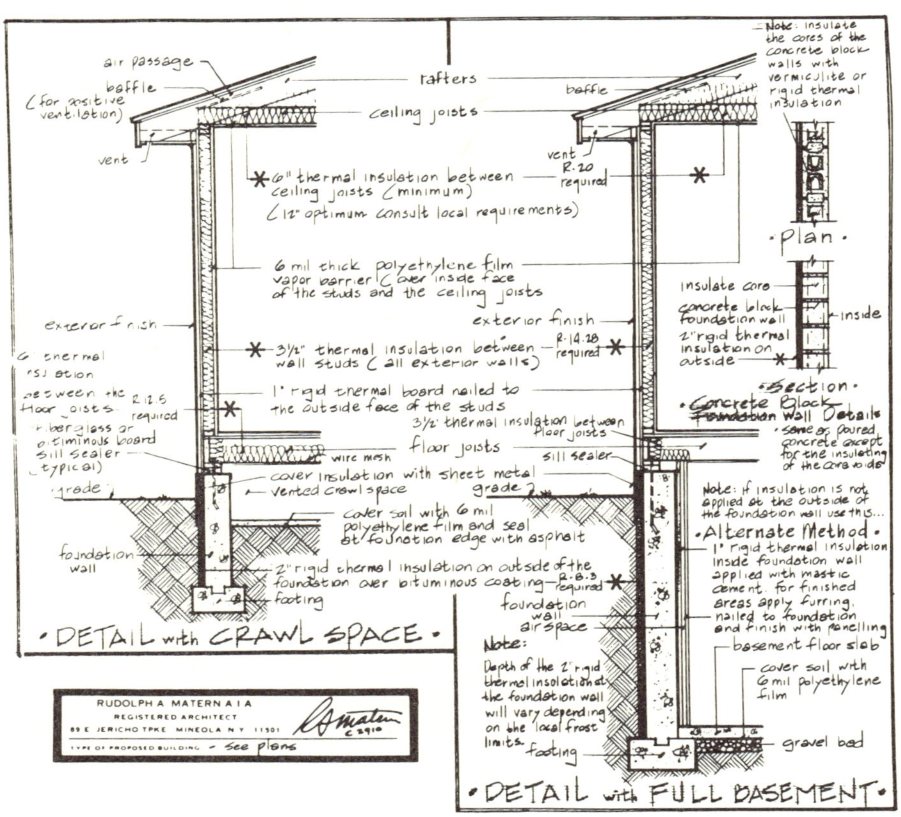

These architect's drawings illustrate the footer and foundation of a home on a crawl space, slab, or full basement. Important insulating details are shown.

Chapter 11

Excavating and Foundations

The foundation walls should be about 12 inches above ground level on the outside of the building. The local building code determines how deep below ground level the foundation will be (usually about 30 to 32 inches). Having the foundation walls 12 inches above ground level allows you to keep your building in line with the general contour of the ground. You wouldn't want your house sticking up out of the ground 4 or 5 feet, nor would you want the floor level to be 4 or 5 feet below ground level. A marker at each side will enable you to keep your house at a correct level.

Watch the machine operator to make sure he puts the dirt from digging your crawl space and footer hole where you want it. You may want all the dirt from the foundation to be put on the side, in the front, in the back, or a little on all four sides. Be sure to tell the machine operator. You should also watch the machine operator so he doesn't knock over your marker stakes.

The foundation hole should be dug at least 3 feet larger than the foundation all the way around, giving you ample room to walk around the foundation. You will need the additional space when building footer forms or laying foundation walls.

After the machine operator has dug your foundation hole, you should have him dig a ditch for your driveway tile. This ditch must be dug between the existing road and where your driveway joins it. You should arrange for your driveway tile and a load of gravel to be delivered at the time the ditch is dug. This enables the machine operator to cover your driveway tile and move any gravel that needs smoothing out.

You should also order several loads of gravel for the base of your driveway at this time. If the driveway base is installed now, you can avoid some heavy rutting caused by tradesmen's trucks and cars. Such ruts and holes can be very deep and unsightly and may even cause troublesome delays if the ground gets soft on rainy days. The machine operator can also smooth out any gravel the truck drivers failed to spread evenly. Some truck drivers can empty their trucks evenly, but others sometimes leave rather large heaps when they start to unload and when they finish dumping the gravel. If you have the machine operator smooth it out for you, it can save you some labor with a shovel.

When your foundation hole has been dug, measure down from the level marks on your corner stakes, and determine what point will be the top of your footer. You can either buy material and build forms, or dig out and pour your footer down in the dirt. Dig the footer like a trench.

Which is best, the trench footer or the formed footer? Some places won't allow you to dig a trench in the earth and pour your footer. The local building regulations in some areas require you to form all your footers. Other places may allow you to form the top half of the footer after digging the other half. It is a little more economical not to form all your footers. But you must abide by the building code in your area. If you dig all or part of it, be sure to keep the bottom flat and the sides straight.

The forms have to be completed before you call for an inspection. When the inspector has approved your forms, you can then pour your footers.

One thing you must remember when you are

pouring the concrete into your footer is to pour your pads for the interior columns or piers which will support the beams under your house. These pads are usually 2 feet square and 1 foot deep. You should also pour the pad for your fireplace at this time. Fireplace foundations are always at least 12 inches deep and should be at least 6 inches larger all the way around than the overall dimensions of the fireplace. If your fireplace is going to be 5 feet wide, the footer should be at least 6 feet wide.

The next phase is the laying of the foundation walls. It is less expensive to lay block instead of building forms and pouring concrete walls.

Should you lay your own block, or have someone lay them for you? Unless you are a masonry craftsman this is one of those times where you need a friend who knows how to lay block. He could lay the block for you, and you could supply the labor needed to bring the block and mortar to him. This may cut the cost of labor for the foundation walls.

This foundation has been poured in easy-to-erect portable concrete forms. Rainwater or melted snow that collects around the properly built foundation at this point in construction is not a problem. →

The top course of block in foundation walls is always a solid block. This helps to keep bugs, spiders, termites, and other crawling things from entering your house. Either use a 4-inch solid cap block or, if it works better, an 8-inch solid termite block.

On the last course of block set the anchor bolts. These bolts are used to bolt your framing and foundation together. Anchor bolts should not be more than 2 feet from the corner either way and should be spaced about every 4 feet on the run of

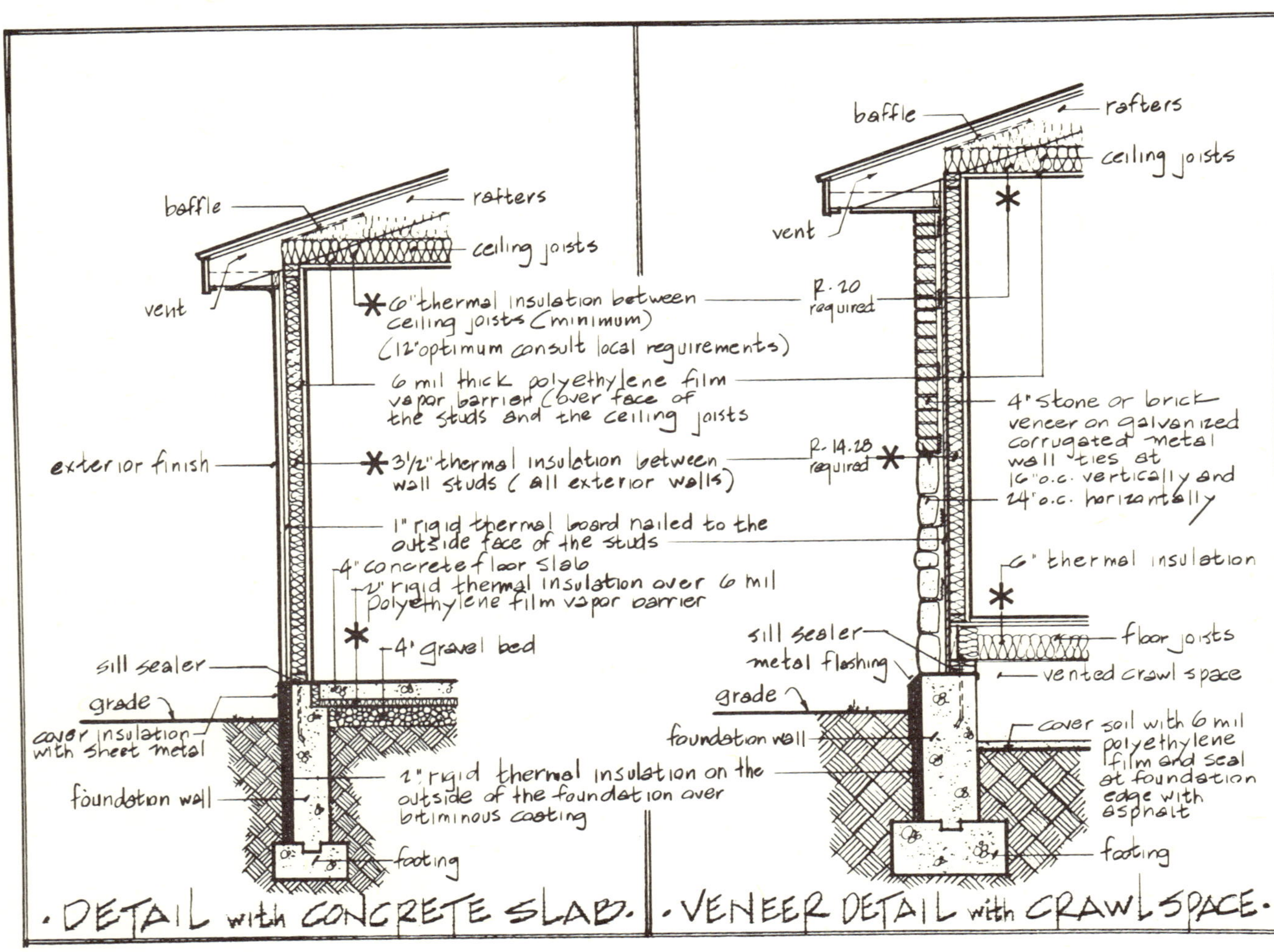

Corrugated window wells prevent low basement windows from becoming blocked with earth, but they have to be cleaned out periodically.

Detail of the Dayton house, window well and vent.

Building permit card MUST be posted in a conspicuous location on the premises and this set of plans MUST be kept on the job at all times. It is unlawful to make any changes or alterations in these plans without written permission from the Building Regulations Department, Montgomery County.

<u>CALL FOR YOUR INSPECTIONS BY PERMIT NUMBER</u>

FOOTER	After footer is excavated and completely cleaned of loose soil and before any concrete is poured.
FOUNDATION (COMPLETE)	Before backfill, after waterproofing is applied.
HEATING - ROUGH	After duct work is completed.
ELECTRIC - ROUGH	Before insulation, lathing or drywall is applied.
FRAMING - ROUGH	After heating, electric and plumbing rough inspections have been approved and before insulation, lathing or drywall is applied.
HEATING (IN SLAB)	Before vapor barrier and wire mesh are placed.
SLAB (LIVING AREA)	Before pouring. Insulation, vapor barrier and wire mesh in place.
FIREPLACE	<u>First</u> - when footer is ready and before concrete is poured. <u>Second</u> - after damper is set, before first flue liner is set. <u>Third</u> - before cap is placed on chimney.
BRICK WORK	<u>First</u> - after tenth course is placed. <u>Second</u> - before last five courses are placed.
LATH	After corner rite and grounds are in place - before brown coat.
FINAL ELECTRIC	When dwelling is ready for occupancy.
FINAL FOR OCCUPANCY	After electrical and heating final inspections have been approved.
<u>NOTE</u>:	It is the responsibility of the contractor or builder to notify the Building Official that a building is ready for final inspection prior to being occupied. Section 115.011 - Montgomery County Building Code.

Calls must be in by 9:00 A.M. on the day inspection is requested; if not, inspections will be made the following day. There will be no inspections made on Saturday.

OFFICE HOURS	8:00 A.M. to 4:30 P.M. Monday through Friday
PHONE	225-4622 or 4623 or 4626 or 4627

A list of inspectors you need in one particular county. Other counties may demand more or fewer inspections.

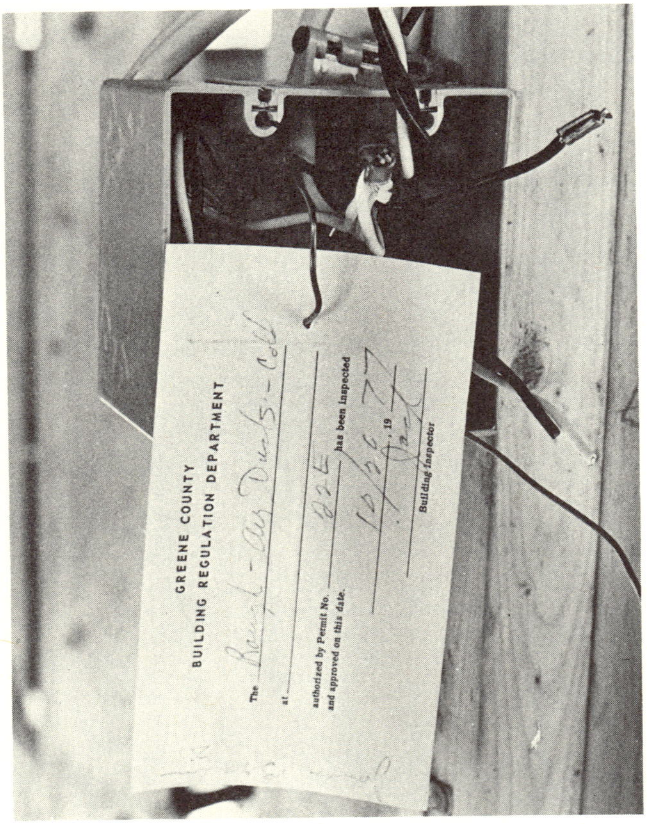

When inspectors approve the various parts of the construction, they usually leave a tag in some obvious spot. You may need the tags to prove to your lender that work is proceeding properly, so don't bury any or seal them up inside the walls.

the foundation wall. The 4-footdistance between anchor bolts works well with block foundation walls.

Foundation vents, which allow aeration of the underside of your house, should equal about 1 percent of the total floor space, and these vents should be installed as the foundations walls are laid. Factory-made vents can be purchased, and they fit into the block work.

This is the area where window wells are used if your vents are down below the finish grade or ground level of your house. Theoretically, if the house has 12 inches exposed above ground, you will not need window wells, but it doesn't always work out that way. Therefore, the window wells will prevent water from running into the house through the vents in the foundation walls.

When laying the block for the foundation walls, also lay the block for the interior columns or piers which support the beam for your house.

Do you call for an inspection after the foundation walls are completed? No, you generally don't have another inspection until you have completed the framing.

Don't worry about getting inspections past inspectors. Inspectors are public servants and are there to look out for your best interest. It may not seem so at times. You may have something in mind that you want to do that doesn't exactly adhere to the building regulations in your area. You could be correct on some point of your

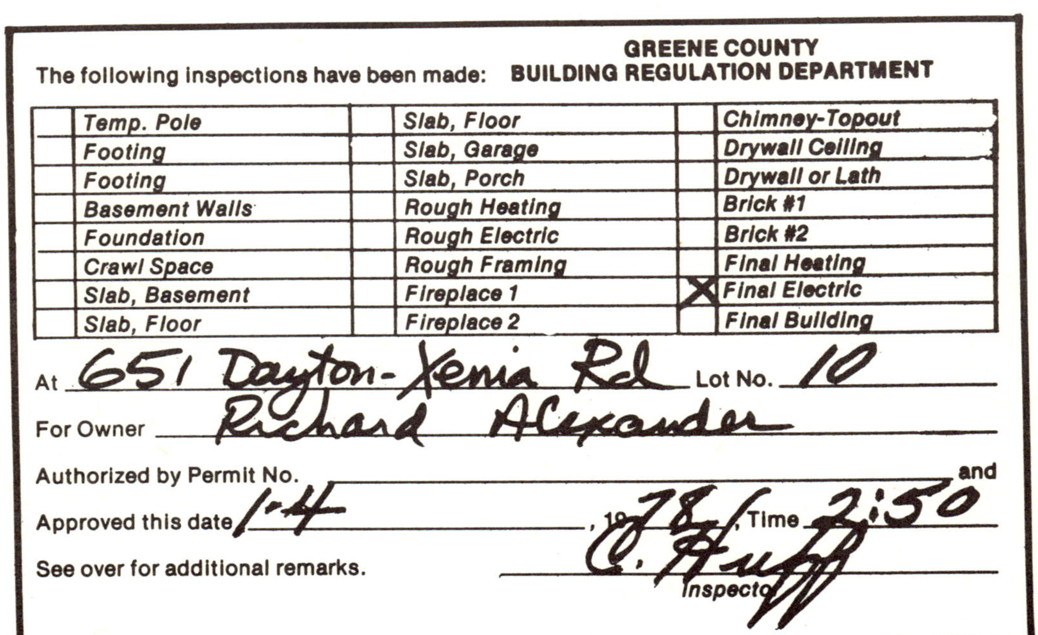

GREENE COUNTY
The following inspections have been made: BUILDING REGULATION DEPARTMENT

	Temp. Pole		Slab, Floor		Chimney-Topout
	Footing		Slab, Garage		Drywall Ceiling
	Footing		Slab, Porch		Drywall or Lath
	Basement Walls		Rough Heating		Brick #1
	Foundation		Rough Electric		Brick #2
	Crawl Space		Rough Framing		Final Heating
	Slab, Basement		Fireplace 1	X	Final Electric
	Slab, Floor		Fireplace 2		Final Building

At 651 Dayton-Xenia Rd Lot No. 10

For Owner Richard Alexander

Authorized by Permit No. ______ and

Approved this date 1-4, 1978 Time 2:50

See over for additional remarks. C. Huff Inspector

Another form of inspection notice, this one for final electrical wiring.

building ideas, and if you explain to the inspector what you want to do, he may be persuaded to go along with you on a particular point. Inspectors are on their jobs to keep you from building something that will fall down, or blow up, or burn down.

Inspectors are also concerned with future residents of the house you build and want to see that nothing unsafe or unhealthy is every likely to occur. You may be willing to risk termite infestation or wall cracks now, but ten or twenty years from now, someone may not want them. The inspectors watch for future hazards, too.

The average family moves every five years. Therefore, when you build that dream house you plan to live in for the rest of your life, remember circumstances change, and the building inspector has to consider who may be the next occupants of the dwelling you construct.

After the foundation walls are laid up, you can then order gravel to cover all the ground inside the foundation walls. The gravel should be spread evenly and be 4 inches deep or more. Gravel serves as a drain for water and stops any vegetation under your house. The gravel also serves as a deterrent for insects and keeps your house from having an earthy smell. Sometimes sheets of plastic are laid on the 4 inches of gravel. Either way, with or without the plastic, there should be 4 inches of gravel under your house.

Call the exterminator to have the building pretreated for termites. The cost of termite treatment at this time is minimal compared to having the house treated after the foundation has been backfilled.

Once the house has been pretreated, call the machine operator to the job site to back fill the foundation walls. This gets rid of some of the dirt that accumulated when the foundation and crawl space hole was dug.

It is best to get rid of the piles of dirt rather than have to carry material around or over them. There is the potential danger that someone will fall down into the hole and hurt himself on the bottom of the foundation wall. It is also difficult to work around the open ditch. So backfill, and this will make for better working conditions. Backfilling at this time also helps the dirt settle around the house as it is being constructed, and when the finish grade has been completed, the dirt will not settle as much.

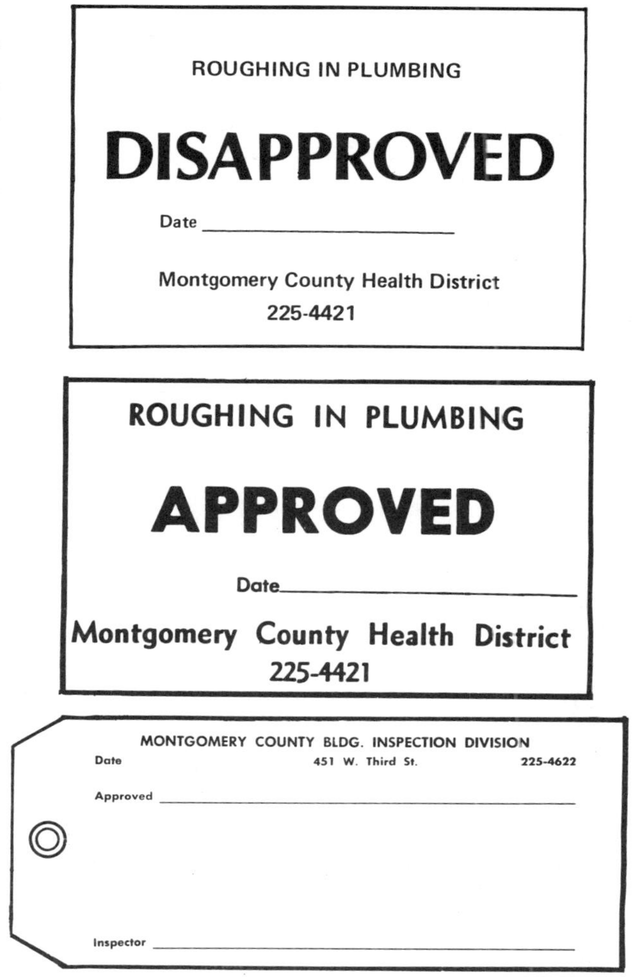

ROUGHING IN PLUMBING

DISAPPROVED

Date

Montgomery County Health District
225-4421

ROUGHING IN PLUMBING

APPROVED

Date

Montgomery County Health District
225-4421

MONTGOMERY COUNTY BLDG. INSPECTION DIVISION
Date 451 W. Third St. 225-4622
Approved
Inspector

These tags are used by building inspectors during construction and may have to be presented to the lending institution.

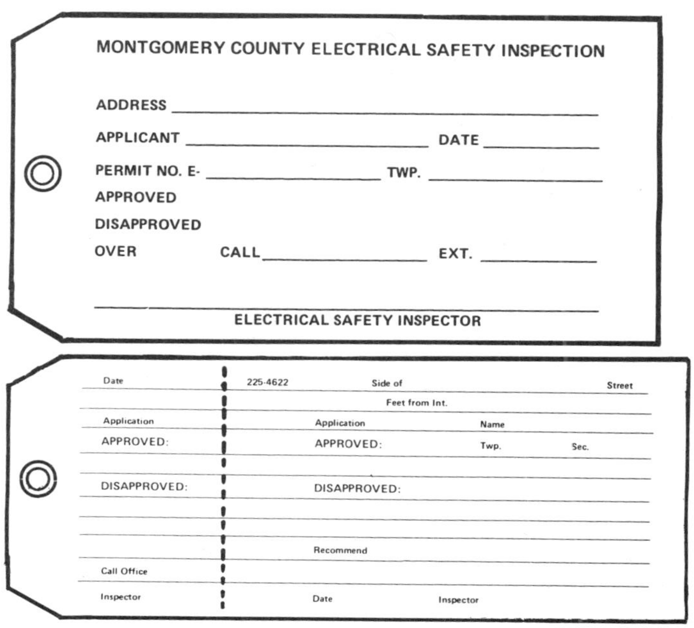

MONTGOMERY COUNTY ELECTRICAL SAFETY INSPECTION
ADDRESS
APPLICANT DATE
PERMIT NO. E- TWP.
APPROVED
DISAPPROVED
OVER CALL EXT.
ELECTRICAL SAFETY INSPECTOR

Date | 225-4622 Side of Street
Feet from Int.
Application | Application Name
APPROVED: | APPROVED: Twp. Sec.
DISAPPROVED: | DISAPPROVED:
Recommend
Call Office
Inspector | Date Inspector

*You can lay carpet directly onto concrete floors,
but for second stories, raised decks, or balconies,
use a particle board called floor underlayment.
It glues and nails directly onto the wood subsurface.*

*Floor underlayment waiting to be cut
and installed in a last patch of wood floor.
In this house,
plywood was used for second-story floors.*

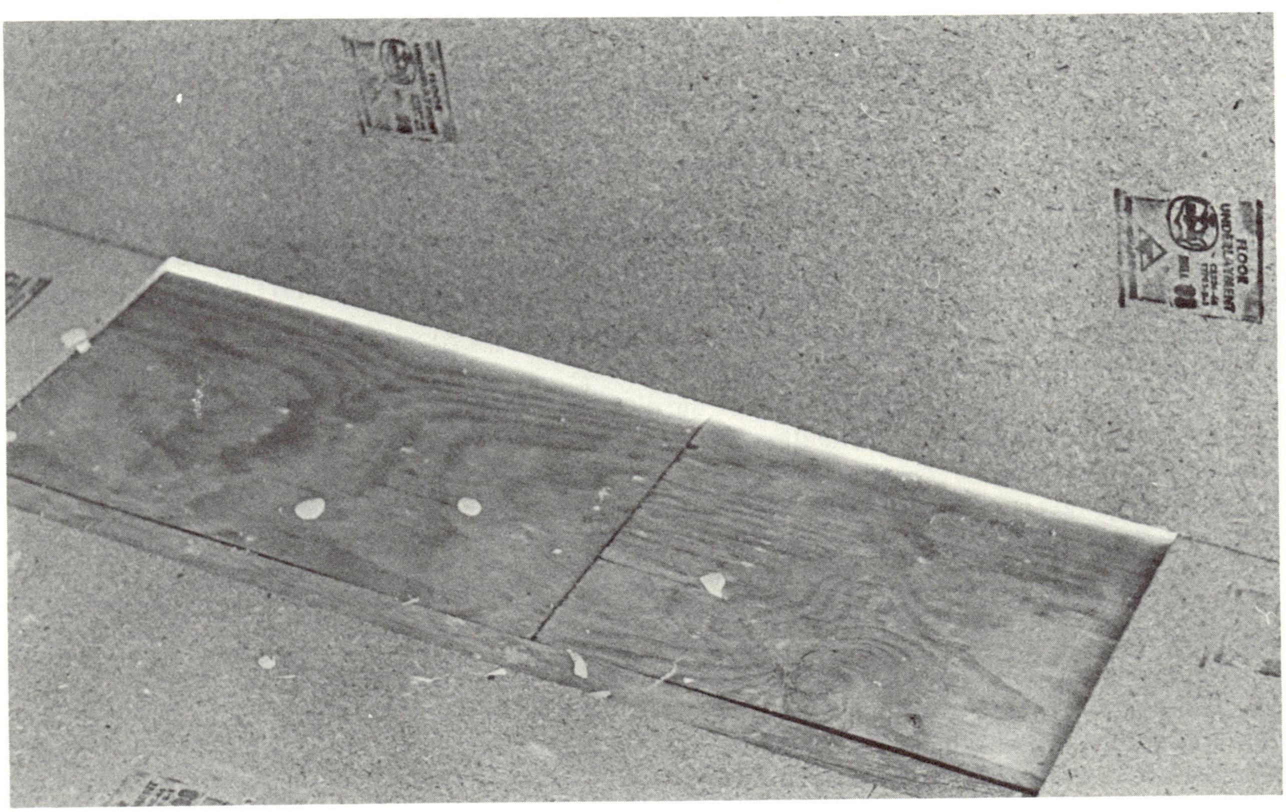

Chapter 12

Building the Deck

The deck, or ground floor, is the next building step. Go by the blueprints when building the deck. If you have someone draw a set of house plans, he should tell you how much and what kind of deck is needed. Have whatever details you need and don't know about included in your house plans. If you are going to have a prefabricating firm supply the material for the deck, the same firm will also supply drawings and instructions that explain in detail how the numbered pieces are put together.

One of the first things, in building the deck, is the beam which is usually made of three 2 x 12s. It should never be anything less than a 2 x 12, no matter what size house you build. The interior columns or piers should be laid out on 10-foot centers. Figure your beam material so everything will be on 10-foot centers. If your house is 50 feet long, all of your material will be 10 feet and 20 feet long. All beam joints should lay on top of the piers, but they should not all break on the same pier. Stagger them out as much as possible.

Check the building regulation code in your locale for the minimum size floor joist for a given span from the edge of the beam back to the edge of the wall. Measure the free span. Free span means the actual open span. Don't count the portion of the span of the joist that sits on the beam or foundation wall.

A typical minimum required joist for a 14-foot span is 2 x 10, but it is not unusual for the free span to be 13.2 feet or something similar. Count halfway if the overall width of the house is 29 feet. When you rule out the thickness of the walls, this could leave the free span less than 14 feet.

Before you start the deck, use an exceptional grade of caulking, or buy a roll of insulation and cut it into strips 4 inches wide. Some other material could be used. Place these as a barrier between the last block of the foundation wall and the first board of the framing. This forms another insulating barrier for your house.

When laying out the joist for the deck use a 2 x 4 plate on top of the foundation wall on which to nail the joist all the way around. The 2 x 4 plates are fastened to the foundation walls by the anchor bolts which were set in the last round of the foundation blocks. Set the beam so it comes up even with the top of your 2 x 4 plates. That makes everything level to build your deck on.

Floor joists are put in on 16-inch centers. All floor joists should be measured from the same wall. When laying out floor joists, be sure to start measuring the 16-inch centers from one end of the house and one end only. Don't make the mistake of measuring 16-inch centers north to south for one end of your joists and then south to north for the other end. If you take a coffee break halfway through the job, be sure to start again where you left off. Don't go to the end and work back to the middle.

Go back and determine exactly where the partition walls will be. Partitions that run the same direction as a floor joist must have at least one extra joist, preferably two, under each of those partitions. Some area codes require two under each wall. Doing this makes your house more solid.

Before putting any of the sub-flooring down, get the bridging in at least every 8 feet. Bridging is used to keep the joists from turning over, to keep them straight, and to tie all the joists together. Bridging can be either 1 x 4 blocks or metal strap bridging fastened to the bottom of one joist and to the top of the next joist in an "x" fashion. Bridging must be used on every space between each joist.

Bridgings should be put in a straight line or row. Take a chalk line and strike a line for your bridging. Again this is to keep your joist from twisting and trying to turn over. Nail into both ends of every block. The bridging also takes some of the bounce out of the floor and makes it more rigid or solid.

Deck material is usually ½-inch plywood, which can be covered with ⅝-inch particle board underlayment. You can also use two layers of ½-inch plywood. To begin, you put down only sub-flooring, which is the ½-inch plywood.

Some people are using a ¾-inch tongue-and-groove plywood as the only flooring. This may not be enough flooring material. There have been incidents where chair legs have gone through the carpet or tile and the ¾-inch plywood. Heavy furniture could cause this type of floor to sag and necessitate having supports later on. The ¾-inch tongue-and-groove plywood is glued to the floor joist and then nailed to it.

Put a nail through the sub-flooring of ½-inch plywood every 8 inches on every floor joist. Some codes state how far to space the nails and how many nails must be used, but most local building regulation codes don't.

Use a #8 common nail. Some people use a #7 coated nail which is easier to drive. But the #7 coated nail doesn't seem to have the holding power of the #8.

Your deck is now finished, and you are ready to move on to the framing of your house.

Keep the larger pieces of scrap in a handy scrap heap, and resist any temptation to have a premature bonfire. There is a constant need for small pieces in odd shapes, and it's cheaper to use scrap than to cut them out of new lumber. Have the fire later, or build a tree house for the kids.

For fine adjustments when tapemeasure calculations turn out a little too big, use a coping saw and pencil.
This tool easily cuts small pieces and can change cutting direction any time.

Chapter 13

Framing

Before you can begin any type of framing, you must determine what kind of windows to use. Different name-brand window manufacturers measure their windows differently. Once you are sure of the brand of windows and know their exact size, you can then determine the rough size openings that you need.

Another thing to consider is the overhang of the roof. This is important because you wouldn't want your soffit return to come back and cover part of your windows. As a general rule, if you position your windows so that you have 12 inches of inside wall space above your windows, this will take care of just about any overhang without getting you into trouble.

After determining all the framing factors of the windows, go back to the blueprints and figure out where your windows and doors will be.

Consult your building regulation code book and determine what size header material—the boards above windows and doors—that you have to use for any given width. 2 x 4s on a barring wall could possibly go up to 3 feet wide. Sometimes it is more convenient to use a 2 x 10 above everything, no matter how wide the openings are, over a double door, a single door, or a double window. That will get a double plate and automatically gives you the right height for the openings. This way you won't have a lot of little blocks to cut, and all your headers will be uniform. In addition, all your windows will be the same distance from the ceiling.

This photo of "framing" illustrates a number of very good points. First, note the 2 x 10 headers over the open areas which will become exterior doors, leading to a sun deck and a fireplace. The doorway, in the far wall, has two 2 x 4s to give an automatic 12-inch clearance between the door and ceiling. Also note the 2 x 4 brace used diagonally, foreground, to hold things upright. Wall studs are 2 x 6 throughout to allow more room for extra insulation.

If you build all the wall yourself, take the top and bottom plates, and lay them out on the deck where they should go. Go through and lay out all the studs on 16-inch centers, using a 100-foot tape all the way across the whole width or length of the house.

If you use a framing square, it may slip back and forth. When the framing square slips, you will gain and lose; then, when other material is applied later, it won't fit. You will have to cut everything. Don't be tempted by a framing square.

Where do you usually start laying out or building the walls, right front corner, left front corner, or on the back wall? It really doesn't matter where you start. Start where you want to. But it is best to build the wall on the ground or floor and then stand it up.

After all the studs are laid out, take the 100-foot tape, and go back through again laying out the center of all the doors and windows that go in the wall. From these center points you can measure in each direction to determine exactly where the cripples and trimmers and all the short pieces go to frame in your windows. Determine

The more an owner can accomplish on his own, the less he must pay someone else to do it for him. Here Jim Brown puts together some rough framing for the kitchen interior.

how long your header pieces need to be after you have the rough openings for your doors and windows. Once the studs are laid out you can start nailing them to the plates.

Determine where the wall should be on the deck, setting it back from the edge a distance equal to the thickness of the insulating sheathing. When you apply the sheathing (½″ for one layer, 1 inch for two layers) it should be flush with the edge of the 2 x 4 plate on top of your foundation. To keep the wall straight, strike a chalk line and set the wall to it.

If the walls are not too long, a couple of your buddies can help you stand up the whole end of the house. If your deck runs a little irregularly, it is possible to make up for it when you put the wall in.

Take a couple of extra 2 x 4 studs, and make braces out of them to hold up the wall until it is nailed in place. Repeat the procedure and get another wall nailed together; then stand it up and brace it. Building walls becomes repetitious.

If you have a long front wall, 50 or 60 feet long, make it up in two or three pieces.

Get all your outside walls up and get braces on them. Put a string line on the inside of the wall close to the top plate, and readjust your braces to make the wall straight. At the same time, make sure that the corners of the house are plumb. Put braces on the corners to hold them straight up and down. Then readjust the braces between the corners to make sure the wall is straight before putting sub-siding on. Your sub-siding of insulation sheathing will hold everything plumb, and after you put it on, you can begin the inside partitions. Braces between the corners, about every 8 feet, should be put back on after the sheathing is installed.

Lay out the inside partitions the same way you

did your outside walls. You can use 24-inch centers on stud work for inside partitions, but it isn't worth the money you save over using 16-inch centers. On a regular ranch-style house using trusses, none of the inside partition walls are bearing any weight of the ceiling, so you can use 2 x 4 headers on them and save several dollars. That offsets the cost of using 16-inch centers for the inside partitions stud work. (Load-bearing walls in non-ranch homes use 2 x 10 headers, and they cost more than twice as much.) In most areas it is not a requirement that you use 16-inch centers, but for the $50 you may save on the overall cost of your house by using 24-inch centers, it isn't worth it. In addition, 24-inch center, inside partition walls are just too wide to try to hang drywall.

After the inside partition walls are set, you should start the trusses. Order the trusses for the roof well in advance. Don't let the house sit a couple of weeks waiting for the trusses to be delivered, but don't order the trusses so far in advance that they are on the job site two weeks before you are ready for them. Just as soon as you finish the inside walls, the trusses should be put on.

It is not advisable to cut your own rafters and ceiling joists unless you are an experienced carpenter. You may save some money, but it will amount to very little considering the time it takes to put the rafters up. It is simply too much work and too much aggravation to cut rafters. You don't save a great deal of money because when you cut rafters yourself, they have to be set on 16-inch centers, and trusses can be set on 24-inch centers.

The many saw cuts on the inside of the 2 x 6 board shown here are to allow bending around a curved surface. Nails will hold it in place. In this board, cuts were made every 1½ inches.

There are many ways to put up trusses and none of them is good. It is all hard work. However, laying out the trusses is rather easy. Start at one end and lay them out every 24 inches from one end to the other. Lay out the trusses on the center line so the roof deck plywood always breaks correctly. Both ends of the plywood will be directly over a 2 x 4, so you won't have to be cutting out pieces of plywood all over the middle of the roof.

While you are putting your trusses up, keep a string on the longest straight wall. Be sure it is straight and as near perfect as you can get it. Your trusses will be made in a gig so they are uniform and have very little variation in them. If the trusses are held exactly right on the back wall you don't have to take all the braces off. Leave the braces loose on the front wall, and adjust the wall to fit the trusses. That way everything will remian lined up, and later on, when you start to apply the outside trim, everything will still be lined up. You won't have to cut part of them off and nail shims to make them look straight. When you get the trusses straight, the plywood sheeting is easier to put on. You won't have any humps in the roof. If a truss is in or out too far from the other trusses, it will cause a hump in the roof.

A couple of furring strips or some extra 2 x 4s should be used when you put your trusses up to keep tacking the point of each one of them together. Put the first one up, and run a couple of braces down to the inside partitions or a longer 2 x 4 to stakes in the ground or to something good and solid. Brace the truss points to each other so they will stand up straight. Measure them at the top where the marks are and on the bottom where you set them on top of the wall. Nail them down, and that will be the top of the house. Measure

The 2 x 6 board that was so carefully saw cut to allow bending is for a rounded porch roof. No board this thick would stand a pronounced curve without those cuts.

A couple of stout boards at the corner of your house provides an excellent semi-permanent scaffold and eliminates the need to carry a ladder hither and yon when working on windows, eaves, etc.

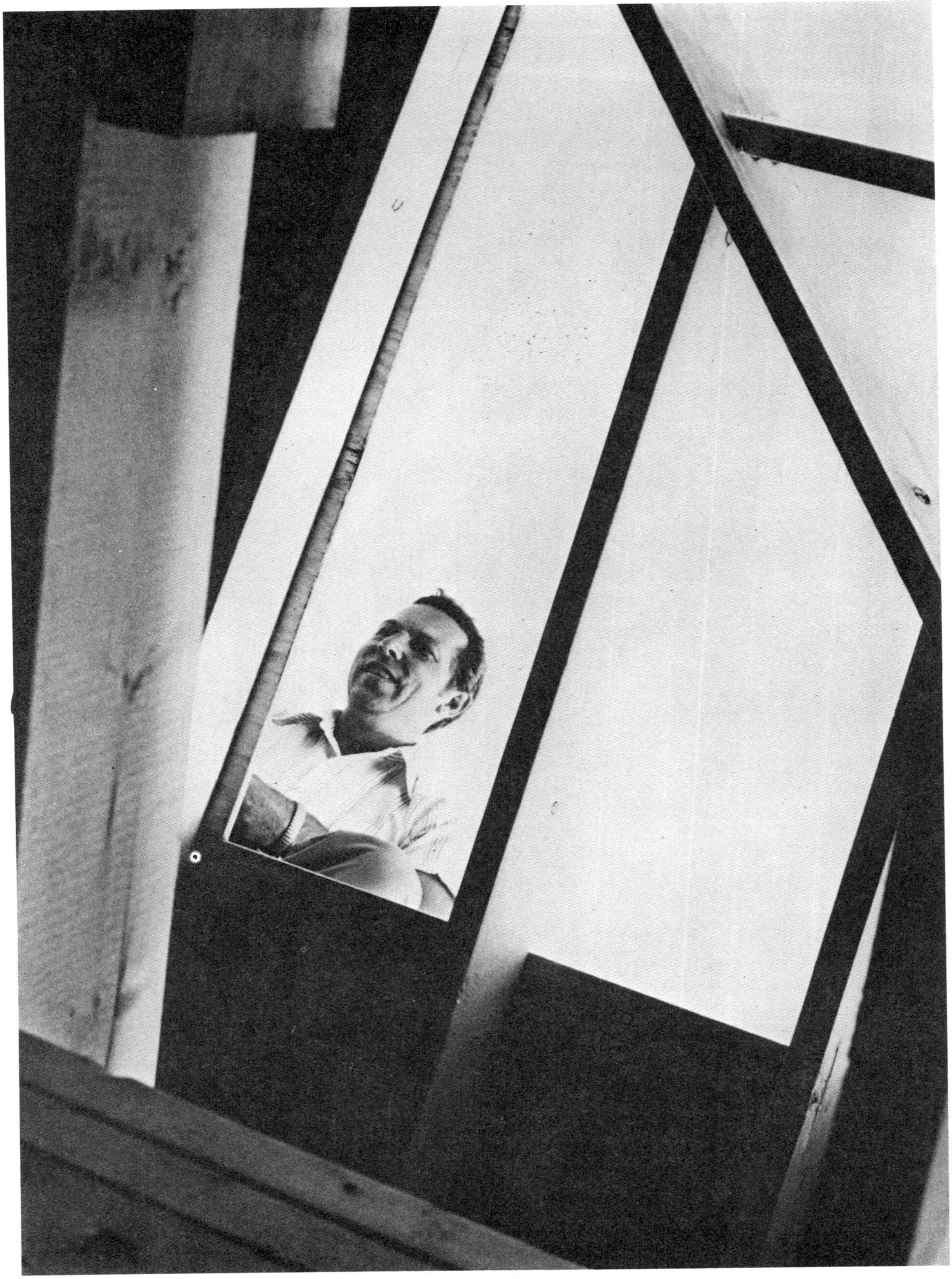

A happy do-it-yourself homebuilder will work faster to close the holes in his roof if he's shy about having his picture taken.

each truss at the top when you set it up to keep it standing straight. Then, when you start nailing the plywood on, it is much easier to keep everything going straight. If you get the trusses crooked, it is a difficult job to go back and pull the nails out and prise them over in place after they have been buried.

After you get the trusses up and straight, you are ready to start putting on the plywood roof deck.

It is best to use ½-inch plywood sheeting, although in some areas ⅜-inch plywood sheeting has been used with the result that after the sheeting has been on for a few years, it settles down between the joints. This makes the roof have little waves in it. To avoid this, use ½-inch plywood sheeting.

Next figure what size fascia board you are going to use and let the plywood overhang the fascia board ¼ inch to ½ inch. The fascia board is usually about ¾-inch thick. Therefore, if you allow ½-inch overhang, that is 1¼ inches. So measure straight up on the run of the rafter 46¾ inches on both ends of the house, and strike a chalk line all the way across the roof of the house from one end to the other. Use this chalk line to line up the first row of plywood. If you are going to have any overhang on the gable ends of your house, usually about 12 or 16 inches, you should already have these attached to the trusses at each end so the plywood can be fitted to them.

Start the first piece of plywood on either end. Start in the corner, and make it as near 8 feet as possible, allowing for the overhang. Then run full sheets of plywood until you get to the other end of the house. Cut whatever size piece is necessary to finish out the first run of plywood.

If you start the first run of sheeting with a 4 x 8 piece of plywood, you should start the second run with a 4 x 4 piece, so the joints won't all be on the same truss. Have the plywood joints break every 2 to 4 feet. Repeat the same thing after you have offset the joints.

If you are skilled, you may want to install your own roofing tile or shingle; if not, hire a tradesman to do it for you. Scaffolding in this picture was put in to make installations of exterior insulation sheets easier.

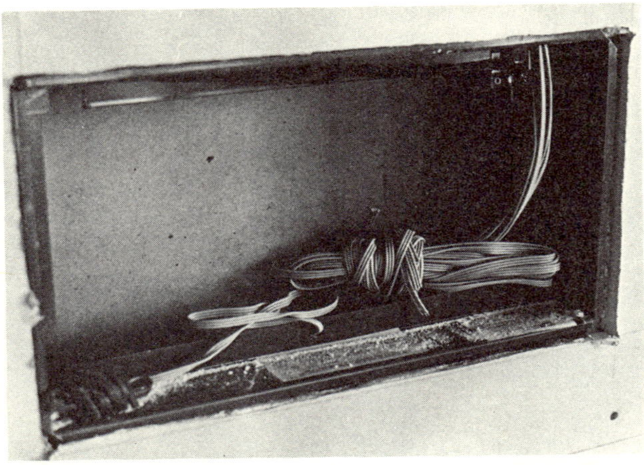

Drywall hangers will be pleased to leave holes for any sort of wiring installation, such as this intercom box, but the time to install the wiring is before they arrive. Intercoms, stereo speakers, ham radio gear, TV antennas, and other items should be planned while the house is still in the frame stage.

This set of roof trusses, sitting forlornly and wet at the side of the road, is testimony to the value of careful planning. The house in which these trusses will eventually be installed was barely in the foundation stage when this photo was taken. They will sit in all kinds of weather for weeks or months until needed and will be warped and difficult to work with because their delivery was premature. A tarpaulin cover would help.

It usually takes four runs of plywood on each side of a house. Repeat the runs over and over until one side of the roof is finished, and then go to the other side.

The plywood sheeting should be nailed every 6 to 8 inches on every truss. Every 2 x 4 should be nailed every 6 to 8 inches. Sometimes, however, the building regulation will specify exactly how far apart the nails should be driven. Don't short-change yourself on nails; use a #8 common nail. Again, some people use #7 coated nails because they are easier to drive and cost less. But they are not as good.

Immediately after you put the plywood on, you should install rolls of black felt paper. The plywood sheeting should never get wet. Use lots of nails when nailing down the felt paper. If the roof shingles are not ready to be put on, and it will be a few days before the job can be done, take pieces of wood and nail them over the joints to keep the wind from getting under the felt paper and blowing it away.

The outside trim: fascia board, doors, windows, and siding (if you are using a siding material for the gable ends of the house) can be installed at this time.

If the job of framing up your own house seems too much of a job for you, shop around for a subcontractor who does only framing work. You supply all the material, and the subcontractor supplies the labor. The charge for framing is usually computed on the square footage of the floor plan.

If you opt for aluminum siding on the non-brick surfaces of your home, installers will create special pieces on the spot to match your house. Here a workman cuts and bends a steel sheet for use as exterior molding.

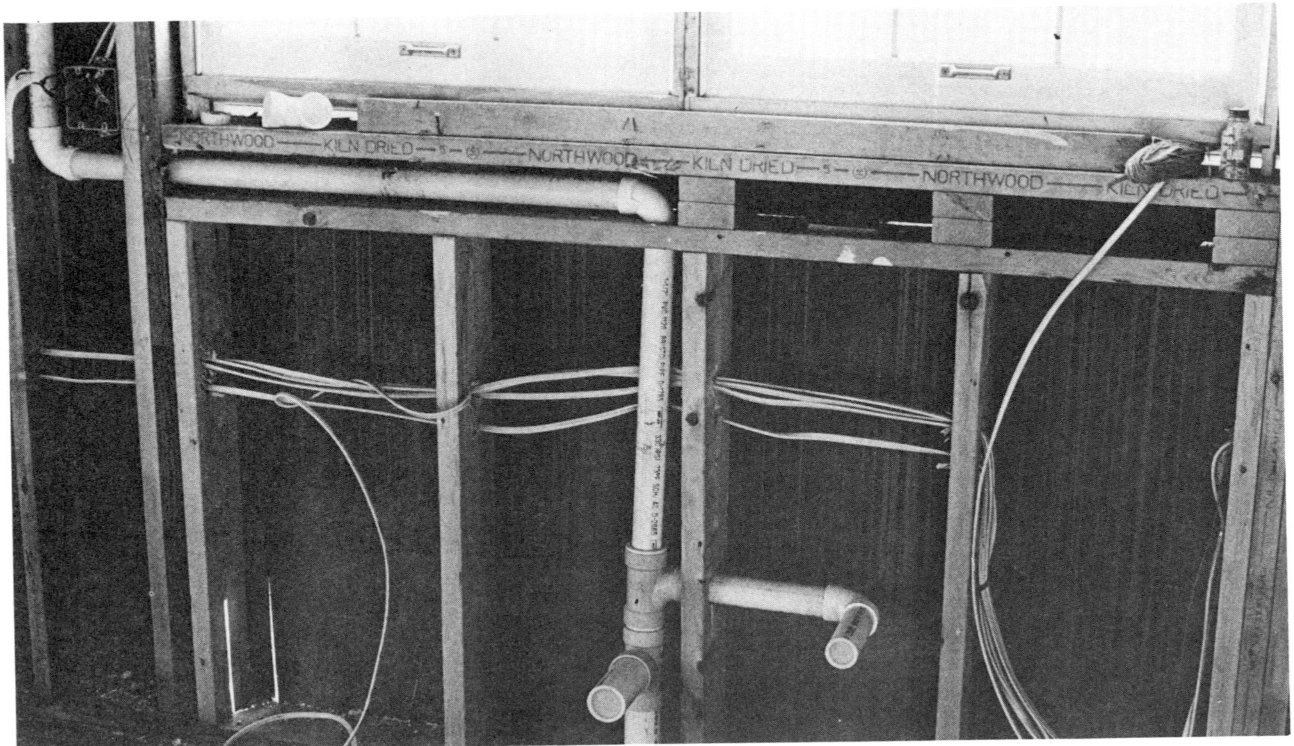

The kitchen drainage pipe, complete with a clean-out installation (lower left outlet) is required by most health departments. Wiring is installed at about the same time as rough plumbing. The loose wire wrapped around its mates at far left will power the garbage disposer. A photo like this in the home's important papers' file will come in very handy if repairs are needed later.

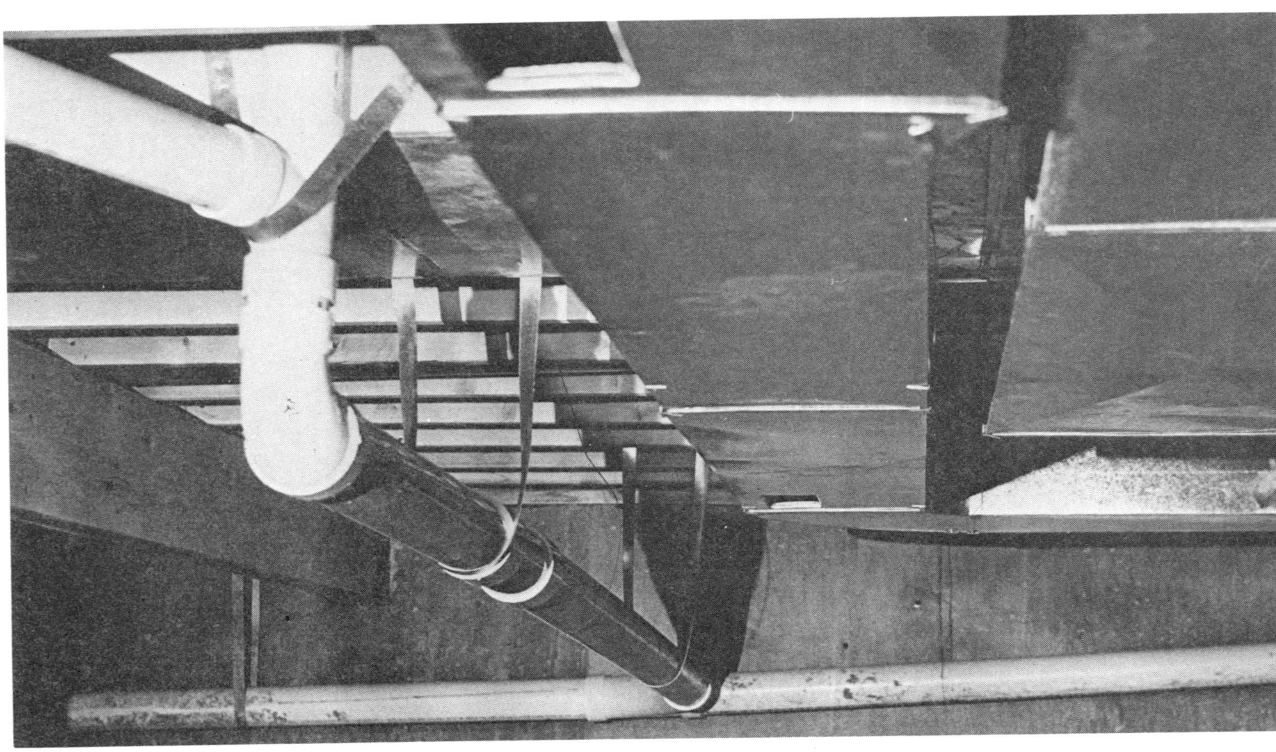

Chapter 14

Rough Plumbing, Heating, and Electrical Work

The job of plumbing, heating, and electrical work on your house should be done by master craftsmen, but you can do the mechanical work if you feel you are qualified to do the job. Most building regulations permit you to do your own mechanical work on your own home if you are going to live in it.

You must have someone who knows what he is doing. This is the time you can use a friend who is an electrician or a plumber to give you some technical advice on how to string your wire or run your plumbing. If you should try to do the electrical wiring or plumbing without the assistance and advice of a knowledgeable friend, you could get yourself into a lot of trouble.

You have come to the point in building where you must keep the whole process moving. There are several different books and pamphlets on how to wire your own home, how to do your own plumbing, and how to put in your own heating system, but they may not meet local code requirements, and this is not the time to sit down and try to figure out what needs to be done. Unless you know what you are doing or have a friend who does, call a subcontractor. You have too much money invested at this point to slow down or stop the building process.

(Left) The outgoing sewage pipes are readily apparent in this basement. They must have enough "pitch," or slant, to carry away waste from kitchen, bath, or laundry areas. This house has a six-inch pitch from right to left. Pipes are hung with strong metal strapping tape from the floor joists. Also visible is some of the heating ductwork. Whether the house has a full basement or just a crawl space, its underside will look much like this.

This is the time to wire or have your subcontractor wire for your intercom, your television antenna, central vacuum cleaner, and any other accessories you may want in your new home. Call the telephone company and have them prewire the house for telephones. This service by the telephone company is free in most areas.

Talk to the subcontractors about all mechanical work. These subcontractors may be able to give you hints that you won't think of. The subcontractors work in their particular fields all the time and are fully versed on all the latest material and styling available. The subcontractor may also be able to call your attention to something you may have overlooked, such as an exhaust fan, a disposal, or a necessary light fixture.

All the subcontractors should be responsible for getting all the required inspections that are related to their work on your house.

At this particular time you can have all your subcontractors working at one time. If you are building a brick house, it is not too early to call your bricklayer and get him started. It is also not unusual to have an electrician, a plumber, and a bricklayer working on your house all at one time. In fact it is preferable. If the subcontractors get in each other's way, they can work it out among themselves on the job and keep going.

At this time the walls are only 2 x 4s, and you can see through the whole house, but you should

This pipe, leading from a prefab septic tank to the drainage field, is thoroughly sealed with pitch to prevent leakage. More sealer is used on the upper, incoming side.

Your septic tank drainage line should be in a ditch deep enough to allow 18 inches clearance to the top. The line, with holes to the lower side, will be covered by layers of gravel, straw, and earth.

walk through the house with your family and look at each room separately. Try to imagine what your daily living involves and if you have provided the necessary requirements. Will the wiring and number of outlets satisfy the furniture arrangement? Is the dishwasher where you want it? Does your family require more lighting or special lighting for hobbies? This is the time to make any changes that may not have been in your plans but are needed. If you wait until later, any changes become major and costly.

After the subcontractors have completed all their work and have secured inspections, you can get your rough framing inspection.

Not many builders or contractors would bother to see that the customer gets credit from the building supply retailer for unused materials, such as this lengthy supply of septic drainage line. The homeowner, acting as his own contractor, will save money by taking it back for a refund, and the supplier will gladly cooperate.

This drainage pipe empties from upstairs bathrooms, drops through the floor, and picks up more pipes from the kitchen and downstairs baths. Plumbing is just one of the many problems to be dealt with in multiple-story homes. This juncture will eventually be concealed by a dummy kitchen cabinet. Plastic pipe is quickly surpassing galvanized drainage pipe because it is generally cheaper and eliminates corrosion problems. Installation of rough plumbing is definitely a job for an expert. It is also one of the important inspection points of interest to county inspectors.

After the electricians have wired the framed structure, but before the utility company turns on the electricity, there will be strange-looking octopuses of wire and cable evident here and there. Don't worry about it. This is the meter box. About 8 inches to the right of the meter box are two telephone wires. Most telephone companies will prewire a home under construction at no extra charge and will obligingly install outlets anywhere the homeowner designates.

This electrical box will house a wall switch or sockets. The two nails holding it to the stud are plainly visible.
Wiring comes shortly after this stage.

When the homebuilder decided where he wanted his lights, sockets, and electrical fixtures, he made an arrow on the ceiling joist.
The electric light box is simply nailed there.
Wiring is run through holes in the joists to another box where a switch will be installed. At this point none of the wiring is "hot," or electrified.

The heater is a maze of wiring that few laymen are equipped to handle; it's best to leave its installation up to the experts.

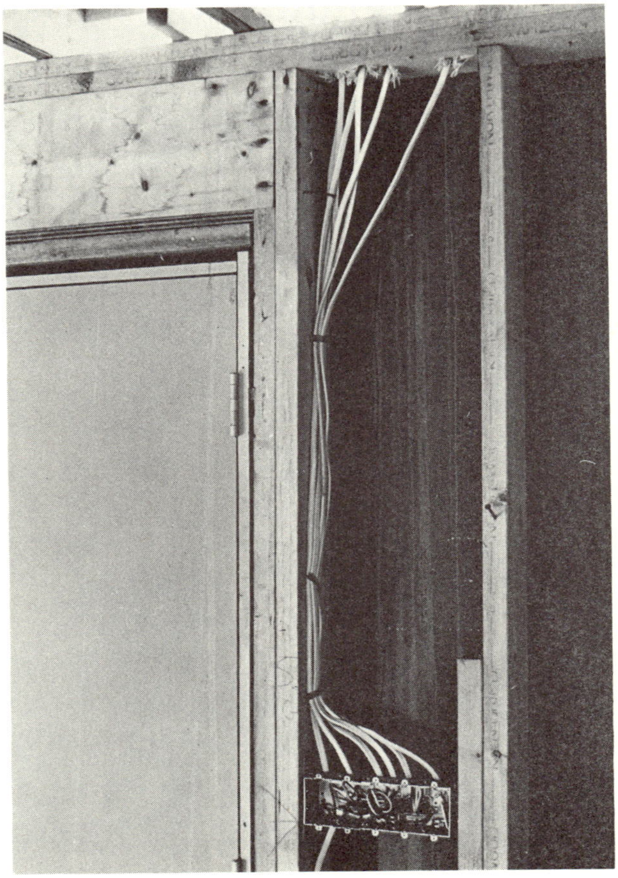

Electricians have installed a five-way electric box next to the front door. Switches will be placed there later for maximum convenience.

A bricklayer begins the first lines of the outer walls. Bricklaying goes surprisingly fast, and then the home really begins to look like the finished product!

When different crews are on a building site together, they have to be careful not to blunder into each other's work. While the owner (with his back to the camera) did some carpentry work, bricklayers and drywall hangers were also on the job here, and even the family pup was watchful.

(Right) A skilled bricklayer takes pride in his work and carefully brushes away all the little stray blobs of mortar that may protrude. The ornamental corner treatment is visible at left.

You can get an idea as to how a particular color brick will look on your home by stacking a few like this. The "shelf" for the outer brick veneer is quite apparent here. It is created by making the foundation wider than the frame walls of the home.

(Right) If the brickwork above this window seems to be sagging a bit, it's because it is! This sort of error is not uncommon when bricklayers move too fast and do not allow mortar to set properly. It is the homeowner's duty to demand a repair job, though, since many tired bricklayers may choose to ignore it. Don't let it be ignored!

How does a bricklayer repair a sagging lintel? By jacking up the wall, as shown, and relaying brick above it. The support should stay in place until the mortar is thoroughly dry and set.

Materials for a fireplace and chimney.

Part of a fireplace with a shelf prepared for a mantelpiece. You can have a mantel of wood, cast concrete, or marble, or even cultured marble which is made of marble dust bonded by cement.

(Left)

Insulation material in batt form easily stays between wall studs by its own friction, but watch out for tiny fibers that fall out of it into your hair during application. Avoid nailing or stapling insulation whenever possible for that compresses it and shortchanges its insulation strengths. It is the air trapped in the insulation that keeps the cold out.

(Right)

Insulation usually holds itself in position by its own tension against wall studs. After installation, this wall is now ready for sealing with plastic sheet and drywalling. For narrow spots between studs, as the one at far left, scraps of insulation do just fine. Since an earlier photo was taken of this piece of wall, the plumber has added a metal sleeve to the kitchen drain pipe, right, and the loose garbage disposer wire has been disentangled from its mates and put where it will be handy for installation.

Chapter 15

Installing the Insulation

Once you have passed your framing inspection, you can begin to install your insulation. Buy insulation to fill the area between the studs. Some people get more insulation space by using 2 x 6 studs instead of 2 x 4s.

Install the insulation between the walls with the vapor barrier to the inside face of the studs and secure the tabs with a staple. Go back and cut all the small pieces above the windows and below the windows.

Cut or tear off small pieces and stuff all cracks, no matter how small. Use a small putty knife, a screwdriver, or other blunt tool to stuff the insulation into cracks around the windows and doors. *Do not leave any cracks without insulation in them.*

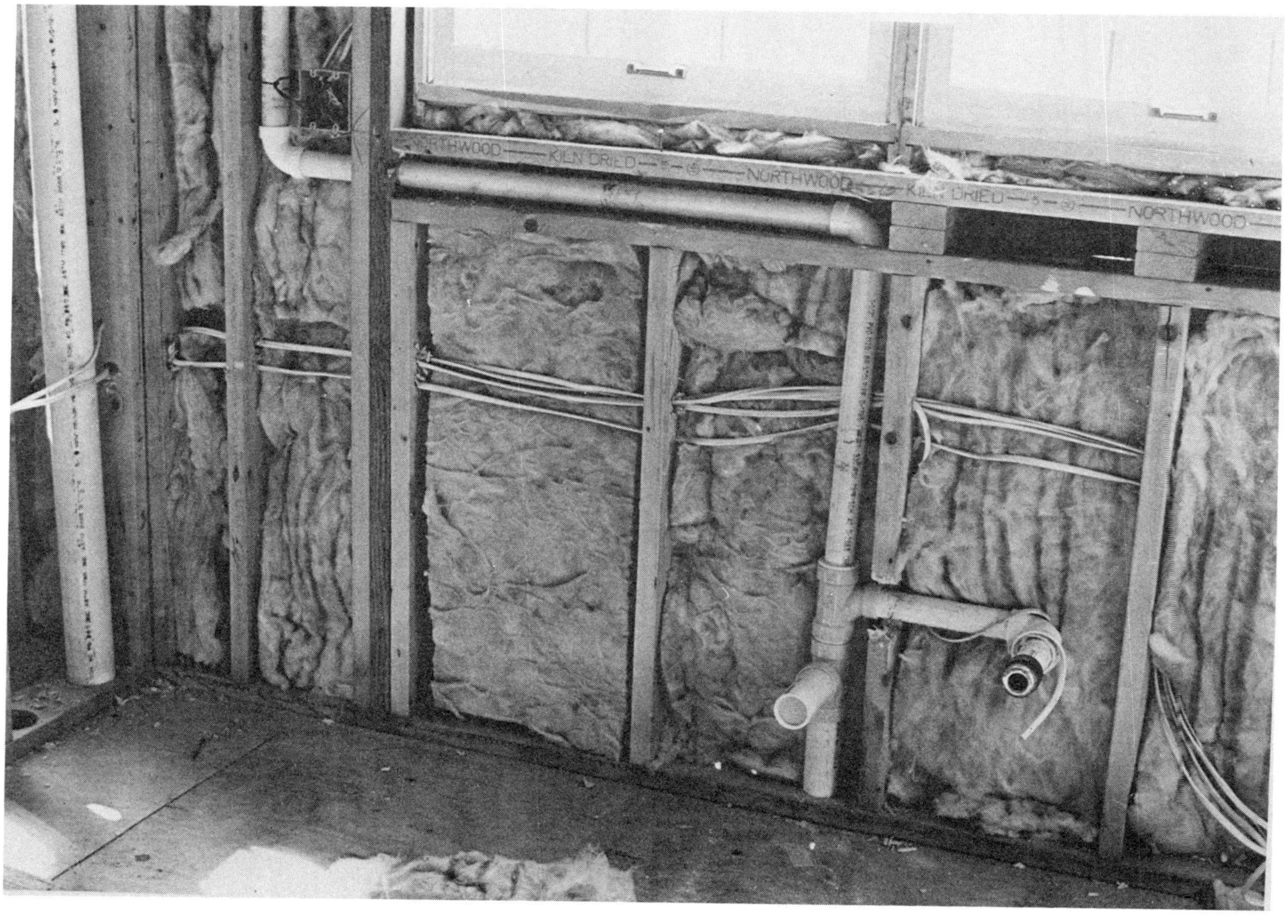

After you have finished, go back and put sheet plastic on all the walls. Buy 10-foot wide rolls of 4 mill or 6 mill plastic. Staple the plastic on all outside walls starting at the floor and allowing the extra 2 feet of width to lap over on the ceiling, stapled to the ceiling joists. When applying the plastic, cover doors, windows, and everything else. Cover the wall in a solid piece. After everything is secured in place, go back with a sharp knife and cut out the openings very neatly. Staple the edges around all the openings.

It is worth the extra money to insulate the walls of your garage, including the wall between the house and the garage. This helps to keep your garage warmer and your house as well. This type of insulation also makes a good sound barrier and can be used in any area where you want to contain sound, such as a garage, a bathroom, or utility room.

Attic insulation is installed after the drywall has been completed, and blown insulation is usually the best way. Blown-in shredded insula-

(Left) Sheets or rolls of heavy-grade plastic cover the insulated walls prior to installing drywall. This is another energy-saving trick, and serves to give an extra seal to the "dead air" in the insulation. The less circulation of air allowed between the walls, the better insulated the home will be. The sheet is installed with a staple gun.

(Right) Help comes in handy when putting up the plastic sheet used in this home as an extra insulation barrier. The plastic is installed with a staple gun.

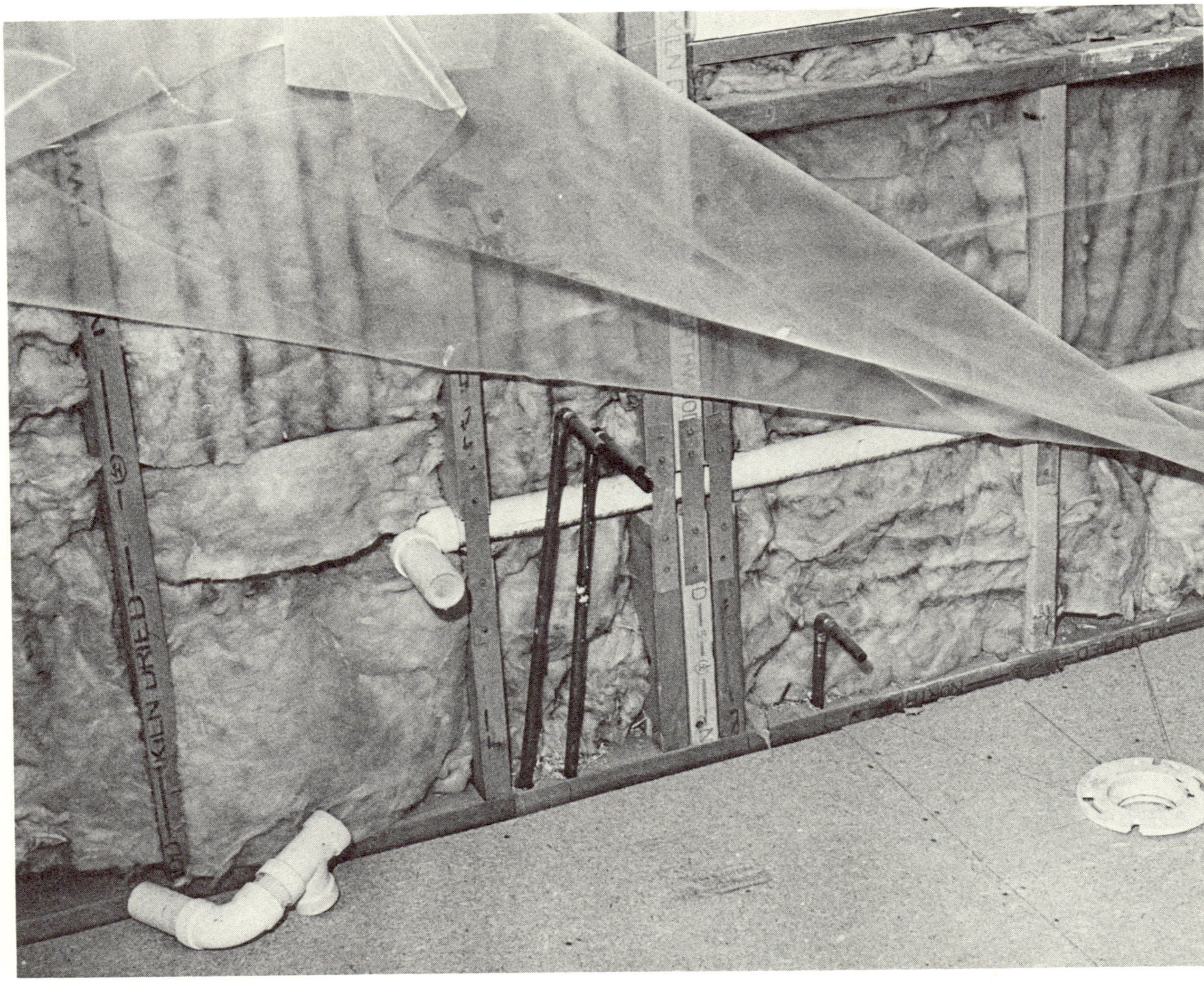

Small dark pipes bring hot or cold water to a sink and toilet to be installed here. Large white plastic pipe is the drain. Fiberglass insulation of this wall is visible. Small pieces of it have been randomly stuffed into crannies because it doesn't have to be pretty to work. Drywall will conceal all this.

tion is more compact, and the forced-air machine distributes the insulation more evenly.

Cellulose and fiberglass are the two kinds of insulation usually used. One is probably as good as the other, but if you choose cellulose, be sure it has been properly fire-treated. Test a handful of it with a match if you have any doubts.

Batten insulation can also be used in your attic. It can be purchased in a variety of widths and thicknesses: 15½ inches wide, 24 inches wide, or almost any width you need. The thicknesses of the insulating batts are usually 3½ inches, 6 inches, 9 inches, and even 12 inches.

The greatest single area of heat loss in your home is through the attic, so don't underestimate the value of insulation. Be sure to use at least the minimum amount recommended for your house. The climate in your area determines how much attic insulation you need.

Electric powered ventilators should also be installed to keep the heat out of your attic. Whether or not you have air conditioning, these ventilators will help cool your whole house.

Black plastic helps water drain away from the walls if it should collect between layers. This photo illustrates simple application of the plastic. Later, layers of insulating material will cover the upper edge and keep drips from getting behind the plastic. This technique is often omitted in arid climates. In humid areas, the bricklayers leave tiny openings in the bottom line of bricks to allow the water to escape.

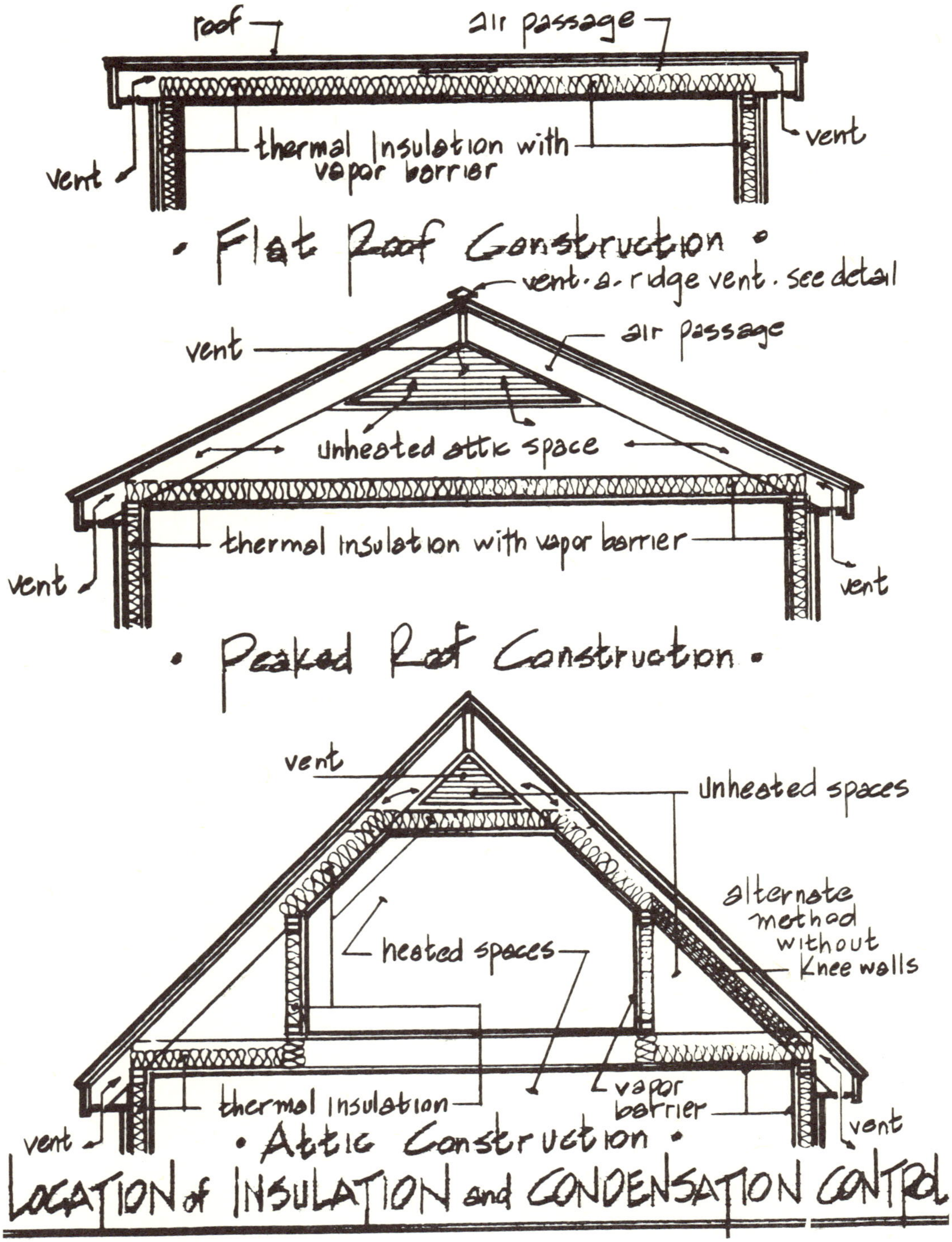

Helpful hints on insulation for an attic from registered architect, Rudolph A. Matern, A.I.A., of Mineola, New York.

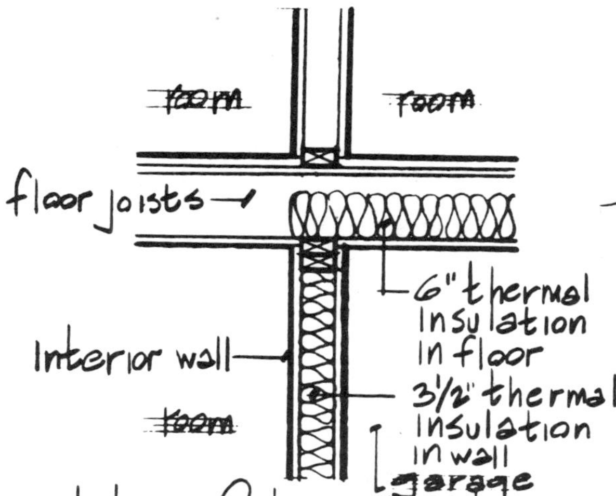

·Insulation Detail at Interior Garage·

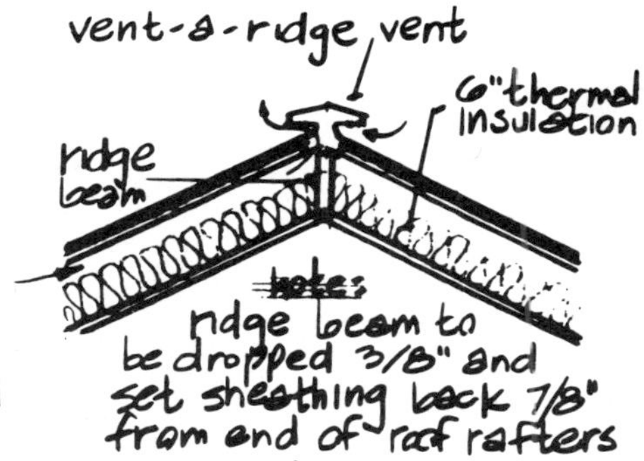

·Ridge Vent Detail·
and insulation of cathedral type ceiling

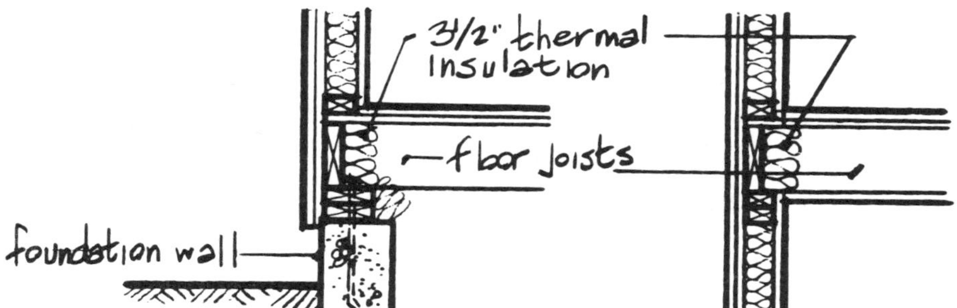

·Band Insulation Detail at Grade and Second Floor·

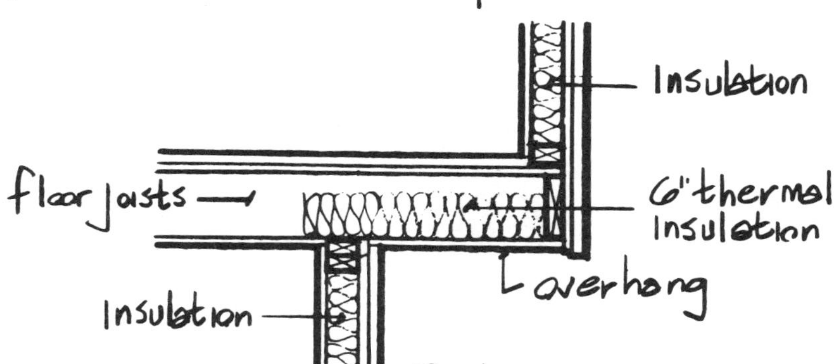

·Typical Cantilever Detail·

·MISCELLANEOUS·DETAILS·

·NOTE from the ARCHITECT·

The information on this sheet is provided to the plan buyer as a service to promote energy conservation and the saving of energy costs through the proper construction of a new home.

Much of the information is common knowledge in the building trade however the new home owner should be aware of it to insure the guide lines are followed and to protect his investment for future years. New national energy saving regulations are now in effect for new homebuilding. Those items are starred * to indicate they are mandatory. The others are recommended for improved energy conservation and should not be overlooked since, for a slight additional cost now, the dollar savings will continue to accrue over the years.

Because of the recent critical energy situation and the reevaluation of techniques some of the instructions on this sheet may show greater requirements than what is shown on the construction blueprints. In that case the information on this sheet should take precedence.

All local building code requirements should be checked, prior to construction.

(Left) Vents in the eaves aerate the attic area. Some commercial builders, especially in dry Western climates, use sheetrock or plasterboard for enclosing the overhang, but humidity or rain eventually causes it to deteriorate. Wherever you build, no matter how dry the climate, use real wood. The savings realized by building your own house allows for the best materials.

(Right) This builder closed his eaves with plywood. Many commercial builders use drywall or gypsum board scraps from inside walls. Wood is definitely better.

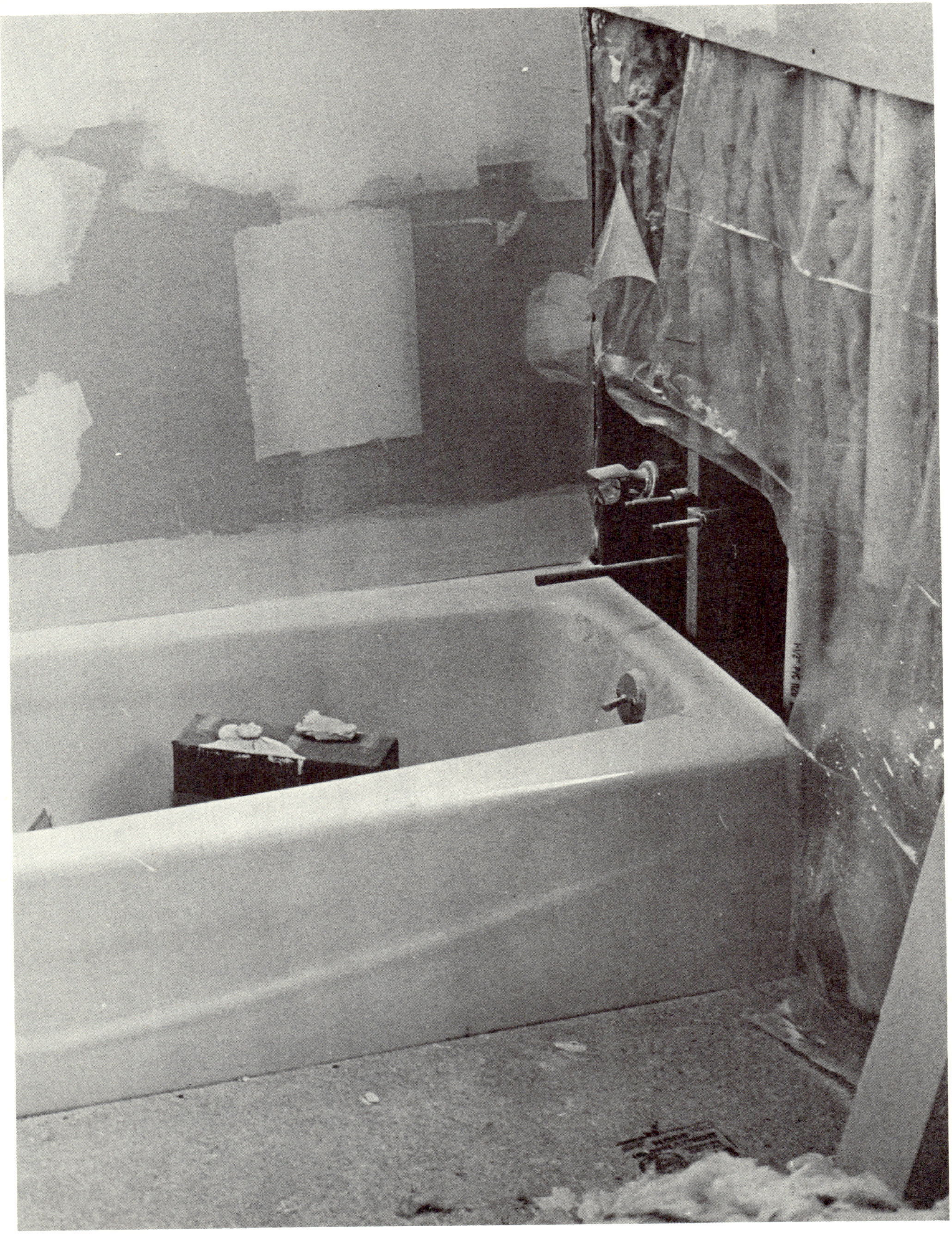

Chapter 16

Installing the Inside Walls

The 4 x 12 pieces of drywall, also called sheetrock, should be installed at right angles to the framing. Never run the sheetrock lengthwise to the framing members you are fastening it to. If your roof trusses are running north and south, then run your sheetrock the long way east and west.

There should be a nail holding the drywall on each end of every framing member and an additional three sets of nails on each framing member. (A set of nails is two nails driven straight in and recessed enough to make a slight dimple in the paper, but not so much that the head of the nail will tear the paper.) It is very important that all nails be driven up so the sheetrock is tight to the framing. The boards should be fitted snugly together but not forced into place.

Large cracks the size of a pencil should be avoided as much as possible. All unnecessary hammer marks, scratches, and broken corners should also be avoided as much as possible. Be sure all joints are made over a framing member.

Try to keep your joints to a minimum. That is why it is preferable to use 4 x 12 sheets of drywall instead of 4 x 8 sheets. Cut the holes around electrical boxes and plumbing lines very neatly. When this is all done you are ready for a spackler.

Bath fixtures should be installed before finishing walls. In this case, the walls are waiting for the plumbers to test all the water lines for leaks. Ordinary drywall is used even in bathtub areas and can be covered with tile later.

Finishing the drywall—filling cracks, joints, and nail "dimples"—is a job for a professional, but if you think you would like to do the job yourself, get ready for a lot of messy work. In addition get someone to help you who knows what he is doing. Now is not the time or place to learn this trade. Remember this is the final finishing of your inside walls before the paint, and paint will not cover up a bad spackling job.

Having the right tools is very important for spackling. Using the correct materials is also important.

There are various joint compounds to choose from. The powdered form is more economical but very messy to mix. The ready-mixed joint compound is more expensive but easier to work with.

For all the first and second applications of joint compound, the all-purpose joint compound is the one preferred, while for the final application the topping compound works best. Using the topping compound on the final application makes the spackling job look whiter.

For the job of spackling a whole house, it is best to purchase all your joint compound in 5-gallon buckets. This is less expensive than purchasing 1-gallon buckets, and the larger buckets are easier to work from. The amount of tape needed for spackling is usually five rolls for every 5-gallon bucket of all-purpose compound you use.

Spackling is usually done in three steps over a period of several days. First, using a 4-inch spackling knife, lay a quantity of spackling compound along all joints between sheets of drywall. Before the compound becomes dry, carefully unroll your spackling tape on top of the compound, and finish with another light layer of compound.

A drywall installer carefully cuts holes in sheets of sheetrock. These must correspond with electric outlets, doors, or windows when the sheet is nailed onto the wall studs.

An opening cut in drywall ceiling for a heating register. The cracks between pieces of drywall will later be filled with spackling and smoothed, with the edges of the opening made neater and, ultimately, a decorative register installed. For now it is enough to have a hole there.

(Right) What does the wall look like after the holes are cut? This sheet of drywall fit perfectly over the doorway and wall switch box. The plastic underneath it is extra insulation and will be trimmed out of electrical boxes later.

(Below) This wall will eventually be concealed by a kitchen cabinet and sinks. Note that an access door was cut from the drywall so pipes and the disposer power can be reached for servicing.

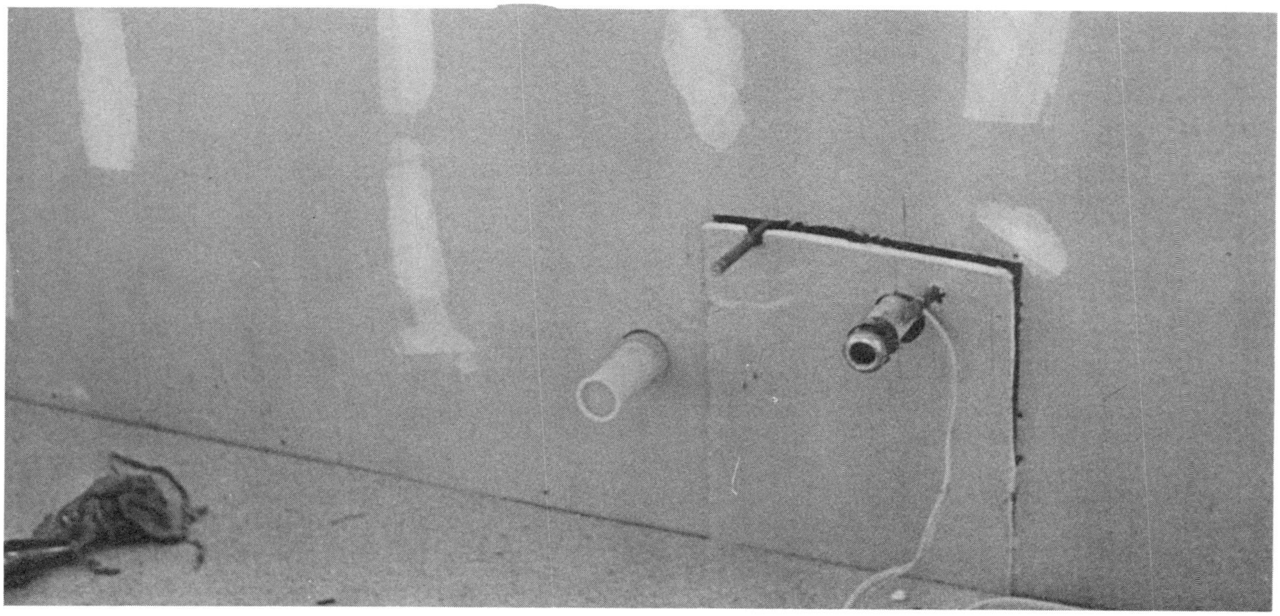

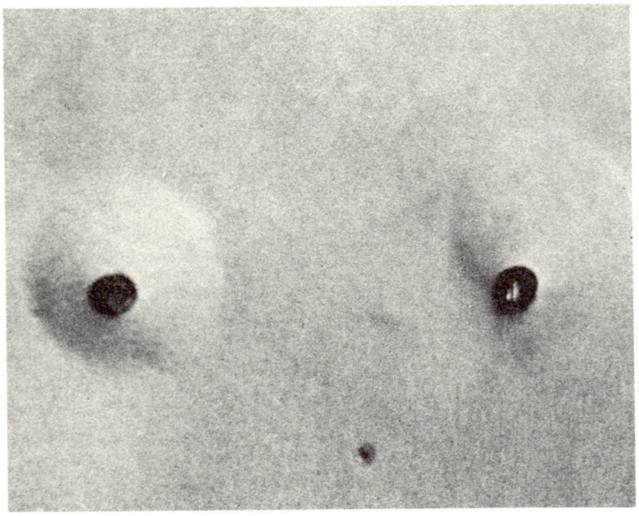

How does a nailhead disappear in plaster drywall material? By being driven a little further into the wall. This process is called "dimpling." The small crater will be filled with spackling compound before painting so the wall looks smooth.

Sharp corners where drywall meets drywall can be a problem as normal wear and tear take their toll. These edges have been protected by a metal cover called a corner bead, nailed directly into the gypsum drywall and then spackled. Corners treated thus are good for the life of the house.

The tools used by a professional spackler are similar to putty knives, but made of thinner metal and more precisely machined – and these spackling knives can be expensive. A do-it-yourselfer can make do with a 6-inch knife and a 12-inch knife, shown at opposite ends of this lineup, and the plastic tub to hold the spackling "mud" while working.

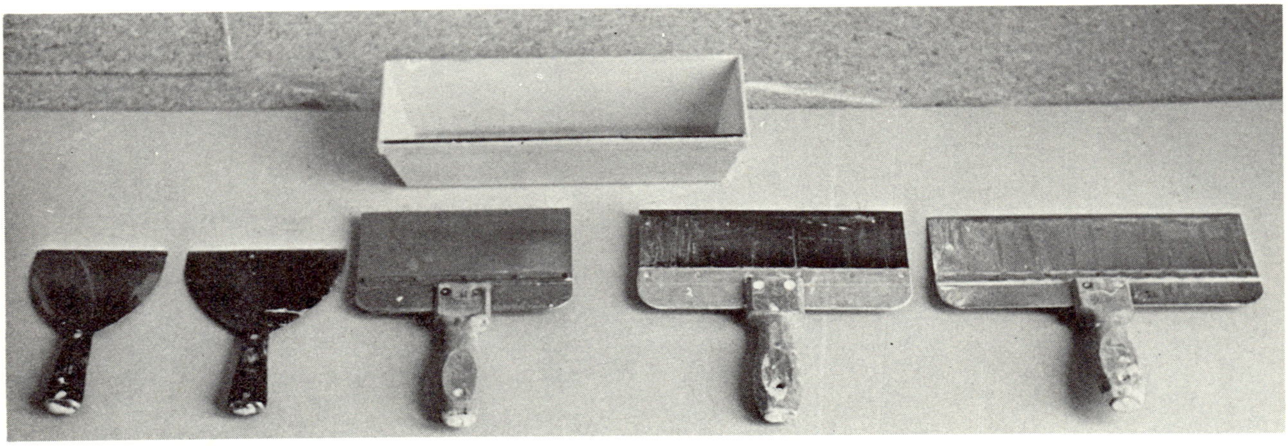

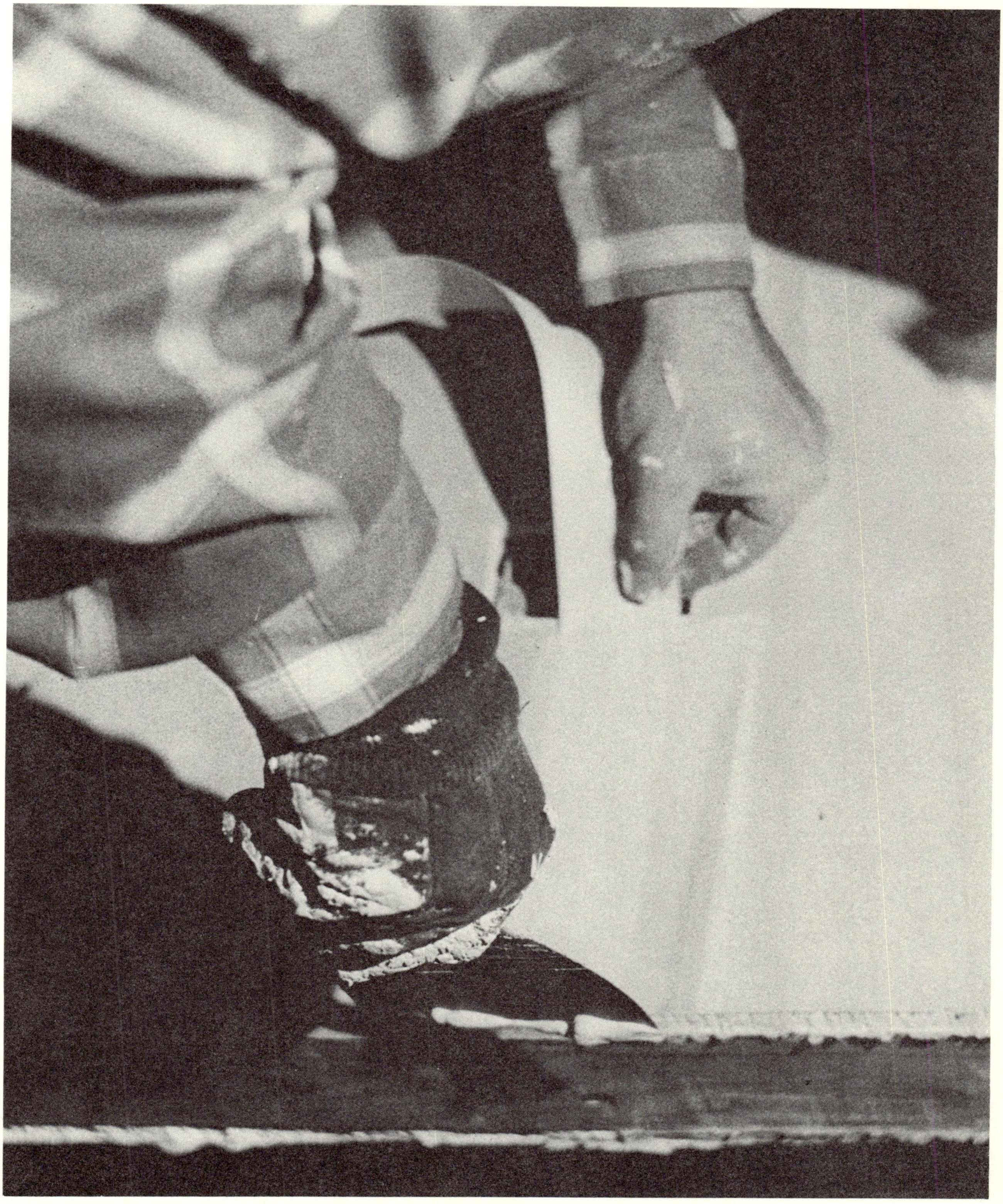

A close-up of the thin paper tape used with spackling to seal cracks in drywall material. It blends well into a wall's surface.

A skilled tradesman applies spackling tape and spackling compound to the crack between two sheets of drywall. He uses a paper tape to cover the cracks, ensuring a single smooth surface on the wall. Spackling compound dries quickly, so the work must be fast and precise. This is not a job for the amateur.

Hold your knife firmly and apply enough pressure so that no air pockets form in the compound.

Don't let spackling compound freeze. If the weather is cold, wait for a warm day, or rent some space heaters to keep the house warm while the compound is being applied and while it dries.

After about twenty-four hours, you can apply a second coat of spackling compound over the first to fill in deep holes and dimples and thoroughly cover all joints in the drywall. Spackling compound is also fine for patching holes that may be accidentally poked in the drywall during home construction.

After the second coat dries, apply the topping compound for the final, smoothest finish. This third layer of compound will cover the tape and all other compound that you have previously applied.

Let it all dry for one more day, and then sand lightly to remove ridges and rough spots. Be on the lookout for marks left by your knife edges. They may look invisible to the fatigued workman or future homeowner, but under a coat of paint they will be very apparent.

Your finished wall is now ready to paint or wallpaper. Some homeowners like to have a spray of a paint-spackling compound mixture put on their walls for a rougher texture. This is attractive if the wall is always kept painted, but consider now whether you are ever likely to paper the wall or panel it. If you think you may want to have something different in the future, put aside ideas of rough textures and leave it smooth.

Paint your walls and ceilings at this time. If you don't know what color of paint you want to use now or don't want to take the time to pick one out, the next best thing to do is paint all your walls an off-white and save the decorating for later.

A ceiling, walls, and corner with cracks in the drywall recently spackled. Dimpled nail holes, where the spackling compound is thicker, dry slower and leave their marks in this first coat. Two or three coats of spackling may be necessary. Paper tape is also visible. It has been folded lengthwise to cover the corners.

(Left) After spackling is completed and dry, paint your new house. Start with the ceiling and work down so spatters won't be a problem. A pole extender for the paint roller makes heights easier to reach.

(Right) Short rollers are adequate for most walls and easily accessible areas. After the ceiling and walls are painted and dry, carpeting will be installed on this wood floor. After the carpet comes the baseboards on the walls.

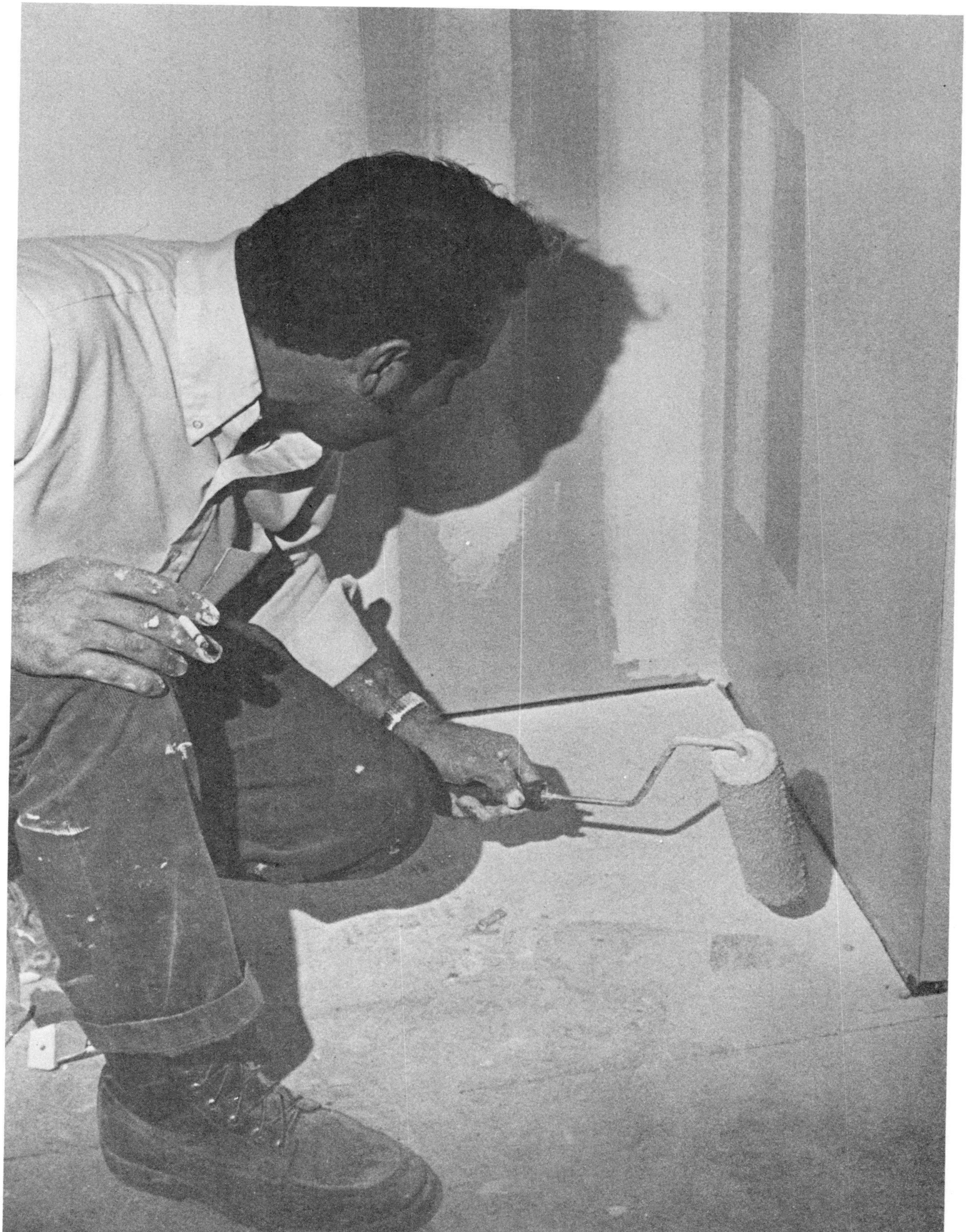

Chapter 17

Interior Trim

Install your floor underlayment, the ⅝-inch particle board, by nailing it to your sub-flooring throughout the house, except where you want hardwood floors.

Install the kitchen cabinets and place the built-in kitchen appliances where you want them. You should know the exact shipping date of the kitchen cabinets and not have them on the job site several weeks before they are installed. This will prevent theft or accidental damage. On the other hand, don't delay your building due to the untimely delivery of your kitchen cabinets.

Floor covering and ceramic tile for the bathrooms should be installed next. When these jobs are completed, call the plumber to finish his work. The plumber will finish his work and get the final plumbing inspection. After the plumber has installed the electrical plumbing fixtures, such as the water heaters, garbage disposal, and dishwasher, call the electrician. The electrician can them complete all of his work and secure a final inspection for the electric work. The heating and air conditioning contractor can come in anytime the plumbing and electrical work is being completed to finish his work and get his final inspection accomplished.

During this same period of time, you can install all your interior doors, the inside trim on your windows, the baseboard, and any other paneling or trim throughout the house. Closet shelves, closet rods, and laundry room shelves are also installed at this time.

After all the other interior trim work has been completed, you are ready for carpet, inlaid linoleum, or vinyl tile if it is used. If hardwood flooring is used, it should be installed before the baseboard and doors are installed.

Any wood trim that is to be stained should be done before it is installed, if possible. Then it can be touched up, varnished, or shellacked after it is installed.

When the power has been turned on in a house, it becomes more convenient to use power tools (some workmen carry portable generators). This plywood is being made into a window seat.

A wall edge in a closet doorway is covered with attractive molding and trim before a sliding door is hung. Nailholes will be filled with putty before painting. Some homeowners might prefer natural wood tones in such trim pieces, in which case they should paint the walls first and add trim later to avoid paint splatters or other accidents.

(Below) A door will hang here, so the roughly cut gypsum drywall stays exposed until special wood trim is installed. No spackling is needed.

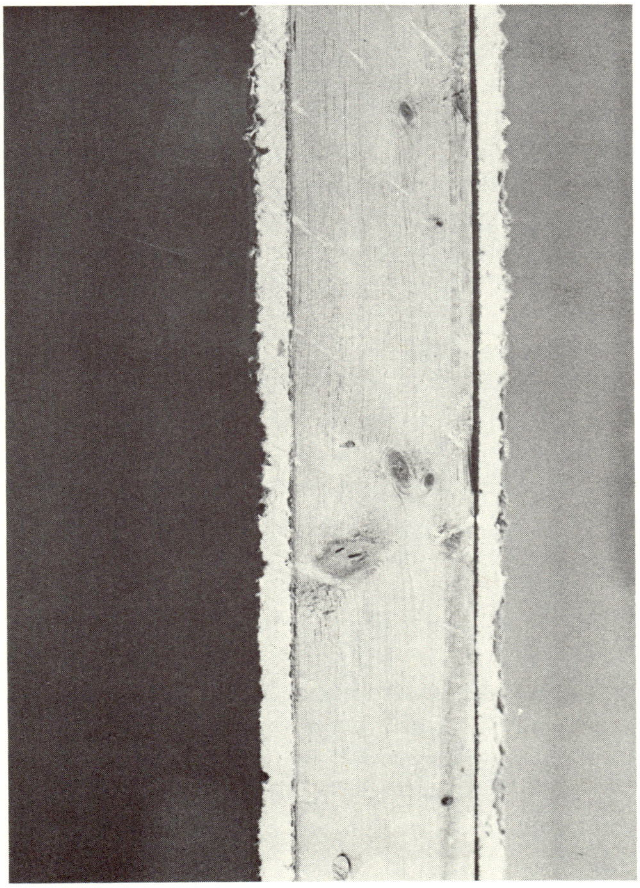

Since the interior trim will be painted, it is being installed before the walls are painted. If window trim, closet doorway trim, doorjambs, or baseboards are to be in natural wood finish, paint the walls first and install trim later.

The stairway is built early during a house's construction, about the same time as the framing, but save the expensive mahogany, teak, or walnut stair treads and risers until finishing time. Workmen's dirty boots, spilled plaster, paint, or other mishaps won't harm the 2 x 4 stair frame during rough construction.

*If your dream house includes a stairway,
you can make the stair treads yourself or buy them
already fabricated in various types of wood.
After installation, cover the valuable hardwood
with cardboard to protect it
from gouging, scratches, or paint spills!*

Chapter 18

Walkways, Patios, and Driveways

Your house is now completed and ready to live in, but there are important exterior details you must now see to in order for it to be attractive, clean, and secure.

Now is a good time to paint your outside trim if it hasn't been done. Don't try to paint the outside trim in bad weather. Be sure to use a good grade of paint so it doesn't run down and streak the rest of the house.

Walkways and the driveway, for example, should be more than graded gravel or stone, since such surfaces quickly disappear with normal use and must be replaced periodically or else leave an unsightly bare dirt path.

Whether you decide on a garage or carport for one, two, or more cars, the thickness of the paving will be more important than the width. Asphalt drives are becoming more popular as a cost-cutter, but you should consider the fact that asphalt or macadam needs regular resurfacing with a protective sealant, or it will break down with exposure to sun, frost, and emissions from vehicles. There is a reason why the garage itself has concrete floors. In the long run, a concrete drive will equally make good sense.

If the house is quite close to the street, where service trucks can park, the principal traffic on the drive itself will probably be the family's ordinary passenger car or light truck. A 4-inch concrete surface, over 4 inches of gravel, will then be adequate.

However, if repairmen, delivery trucks, or other heavy vehicles, such as house trailers or motor homes, will be using the drive, invest in an extra 2 inches of concrete, and protect the driveway from cracks and deterioration. A small crack now, especially in a part of the country where winter brings hard freezes, fills with water during each rain or snow, leading to erosion or freeze damage.

You want the driveway flush with the surface of the ground around it, so excavation will be required. Drainage away from the house must be provided for. Then wood forms have to be securely constructed so that the heavy concrete, when poured between the forms, will stay where you want it and not leave you with a messy clean-up chore.

The driveway keeps the family vehicle from tracking mud into the garage. Now you must see that walkways do the same to protect floors and carpets, and keep mud off the family's shoes.

In the front of the house, walkways are traditionally made of concrete 4 inches thick, over 4 inches of gravel, to accommodate not only pedestrians coming and going, but heavy furniture and appliances that will be delivered or carried out from time to time throughout the life of your home. Construction of a concrete walk is much the same as the driveway although on a much smaller scale.

If you are handy with a trowel and float, you can design you own individual walkway surface with radiating designs in your expansion cracks, or use inset stones or other decor. As in all phases of construction, it is best to avoid too much individualization that forever brands your home with something that future owners would not choose. Resist the temptation to mount a plaque with the family name in it or to carve your initials in bold letters across the walk.

A sensible design item is a cross-hatching of

It's never too soon
to paint the attractive front doors.

lines close enough together to help pedestrians gain traction on rainy or icy days. Make sure that all surfaces drain off the walk so that water will not collect there. Also be aware of rain runoff from the house itself so that visitors or family members will not be forced to walk through torrents. If the walk is to be covered by a porch roof, be sure the roof extends far enough to cover the walk and all who walk on it.

In the back- and side-yards, walkways will not have the heavy traffic of the front entrance, so thin flagstones or decorative brick patterns may be adequate. These surfaces have the additional advantage of being relatively portable, so you can take up stones or bricks and move the pathway as time passes and your needs change. Gravel or small stones could be adequate for garden walks, but keep in mind that they will be uncomfortable to bare feet if you anticipate a swimming pool or like to walk barefooted. In addition, your children may like to throw small stones and may eventually heave your walk over the neighbor's fence.

If you plan to have a backyard garden, and the climate in your area requires regular hosing of the plot, watch where the water drains, and consider paving that area with stones or brick, especially if that is where you will walk when watering.

Flagstones or brick allow grass to grow between the artificial surfaces, so be sure the stones are mounted flush with the earth. This allows you to pass through with the lawnmower and avoid laborious hand-clipping of every blade of grass.

A patio allows comfortable fair-weather activity outdoors, so plan yours before moving into the house. It should be near a door leading to the kitchen, although some homeowners find that bedroom entrances are also desirable, especially if a pool is planned. But most likely, you will be having barbecues on your patio and may not want to carry dripping plates of greasy spareribs over the bedroom carpet.

An important consideration is that the patio is a large, hard, smooth surface that will allow rain to run off. Where will that water go? Be sure the patio is graded away from the house, or you will have puddles against the foundation.

Hopefully, you provided exterior electrical out-

A detail of a set of forms for a driveway shows how the stakes support the lateral boards. Don't make the forms too high, or it will be difficult to trowel the pavement level. Forms can be removed as soon as the concrete is thoroughly leveled and set.

In suburban areas where a driveway must cross a drainage ditch, you will probably be required to install a drainage tile to insure free water flow.

lets and a telephone jack for your patio when those items were being installed in the rest of the house. See that they are finished now. Make sure that electric outlets are equipped with protective self-closing doors, and ground fault circuit breakers, to protect you from being electrocuted in rainy weather.

If your patio is directly outside the kitchen, perhaps a two-way cupboard with a lock on it will be handy now. You can pass food and drinks through it.

Not all patios have to be concrete, of course. On this surface, you can use your imagination. Redwood, tiles, brick, outdoor carpet, stone, or other decorative flooring are comfortable and attractive, and the patio is one spot where a little roughness in construction will not offend but, indeed, even add to the rustic charm of outdoor living.

If you use a wood flooring, it will probably be mounted on small pilings and elevated above the surface of the ground. Be sure, then, to provide access so that dropped eating utensils or other small items can be recovered easily. Also, make sure that the barbecue pit or other fire source won't threaten conflagration.

If you have leftover bricks, stones, or concrete, place them around exterior water faucets where drips are likely to make muddy spots. Four 2 x 4 scraps easily make an excellent form for that extra concrete. Make a hand- or footprint in the resulting concrete tile, write your name and date in it, or let a child make a footprint, and place it under the outdoor spigot or downspout.

If your home is on a newly graded lot without mature trees, you can mount two 4 x 4 timbers in

Before a driveway can be built on the soft loamy soil shown here, a sub-surface of stone must be spread to prevent sinking under the weight of vehicles. This drive will probably be asphalt.

Cim (CQ) Loose of Loose Nursery and Garden Center in Dayton shows some composite concrete tiles for use as garden borders or walkway paving. They can be set into mortar or simply buried at the soil's level. Round, square, rectangular, and free-form tiles are available at retail outlets throughout the country. Most are concrete rather than stone.

12 inches of leftover concrete, screw in hammock hooks, and hang a hammock there. Make the foundation holes 2 feet deep and 10 inches wide, and use a quality wood sealer on the underground portion of the post. After shoveling the concrete around the post, use the pointed end of your trowel to spear the concrete and fill any air pockets that may have developed. Use a carpenter's level to be sure the posts are upright.

This is a good time to purchase a caulking gun and several tubes of caulking to caulk all cracks, no matter how small. Cracks around windows, doors, soffit, and gable ends should be caulked, as well as any cracks that could let air into your house. It is extremely important that all cracks be sealed. Use one of the new and improved weather seals to keep wind from blowing in around your garage door.

This concrete driveway is too narrow, a condition which has led to rutting of the lawn at the pavement's corners. Even if the homeowner learns to steer properly, motorists who use the drive to turn around will cause this problem. Be sure your driveway is wide enough.

A combination of surfaces: solid concrete sidewalk leads to the front door and around the house; round concrete tiles make a decorative walk in the other direction; part of the lawn is covered with ground stone (foreground). Note the evergreen border planting in the background — expensive, but attractive. The ornate mailbox came from an ornamental iron shop.

Making concrete forms is one of the uses you will find for your valuable scrap pile. Here, a patio extension is being created by pouring in 4" or more of concrete, smoothing it, and letting it set. When you build forms, a line level or long carpenter's level helps keep the tops of the lateral boards even with existing pavement.

Chapter 19

Odds and Ends

Now is the time to call the machine operator with a small dozer to come out to the job site and finish grading your lot. This should not be done until all the outside concrete has been poured, including the driveway.

Window wells and splash blocks can be installed after the finish grade. This is also the time to call for a final building inspection.

Now you can go inside your new home and shut the doors and windows when it rains, except that the rain draining off the house will turn the whole yard into a sea of mud. For this reason, you need rain gutters and downspouts.

Guttering and downspouts can be subcontracted for about the same price as you can buy the material for the job. Therefore, it is to your advantage to call the subcontractor and have him install your guttering and downspouts.

The gutters, which come in handy lengths at your hardware store, if you insist on doing it yourself, should be hung along the eaves of the house, using special long nails and tubular nail holders. Place the tubular holder inside the gutter, drive the nail through the outside of the gutter, through the tube and the inside (house side) of the gutter and into the wooden surface of the roof edge. Gutters can easily be cut and joined together for short or long rooflines, so every bit of roof that drains to the ground can be covered, and your garden, lawn, and foundation can be protected from runoff.

When joining extra-long lengths of gutters, use the special waterproof cement available where gutters are sold, and be sure to support the gutter with a nail or two near the juncture. If the gutter is too long, the volume of water that reaches the downspout will overpower it and cause backup and overflow, and don't forget about the weight of winter snow either, so put in enough downspouts, too.

Downspouts are usually installed where the gutters end and easily fit over the small tube openings in the floor of the gutters. Brace the downspouts with brackets or sheet-metal holders which can be affixed to the exterior wall with masonry nails. Elbows and short lengths of spouting will bring the runoff water from the gutter to the downspout if your home has wide overhangs. Another elbow and perhaps a short length of spouting will carry the water away from the house near the ground. Rubber, stone, or concrete tiles help prevent erosion where the water exits.

In the far western deserts, gutters and downspouts seem a bit superfluous since there is so little rain, but they nevertheless go a long way toward protecting the foundation and eaves from water damage and are recommended wherever the grading of the soil does not carry rainwater away quickly. Gravel or wood chip groundcover help prevent splashing mud wherever downspouts are not used.

Everywhere else in the country, gutters and downspouts are strongly recommended on every roof edge.

Rubber tubing can sometimes be affixed to the bottom of a downspout and effectively carry water several feet away from the house. Some such tubing is equipped with a spring wire skeleton which automatically rolls the tube back up when the rain stops. But this roll-back is effective only where the ground is evenly graded and

Chain-link fences are quite strong but not very handsome. This one has been woven with metal slats for more privacy, but slats must be replaced periodically. In some areas you may be required to use chain-link fencing if you want a swimming pool.

all the water is evacuated. If pockets of water form inside the tube, it will have to be emptied manually.

Another detail of a new home is a fence. Fencing is definitely a feature in a house's favor when it is time for resale and is a good investment in terms of security and privacy, too. In many locales, a fence is the difference between having to keep a dog leashed or allowing it to run in the yard.

Cedar, redwood, or other lumbers, pole, or picket are several fence types. Brick or stone walls are most effective but very expensive since they need a concrete foundation and require special skill to erect them.

Chain-link fences are also effective, but do not afford full privacy and require stretching tools and pipefitter's skills to erect correctly. Chain-link is, nonetheless, very popular since it is a hardware item and sometimes offered at good discounts.

Lumber fencing gives fuller privacy if it is used to full height, and the boards are close enough together. Wood can discolor with time, however. The boards are nailed to 2 x 4 or even 2 x 2 crossbars mounted between 4 x 4 posts. If you decide on a wood fence, specify galvanized nails for extra grip on the wood and prevention of rust, or a

A short fence made of vertical boards is decorative and effective in keeping pets or children inside a yard, but for privacy it's usefulness is limited.

A stone wall is sturdy, attractive, and stately — but very expensive!
In the background is a plank fence, useful as a decorative accent or a marker for a boundary.

heavy wind may someday take your fence away.

Pole fences and rail fences are the types usually seen at horse ranches and serve only to decorate or define property lines in a residential setting. Horses may be confined by rail fence, but children and small animals, such as dogs, quickly learn to ignore them.

A length of rustic rail fencing makes an attractive decoration on a lawn and looks better as the wood weathers. Ultimately, however, such a fence will have to be replaced when the rails rot away.

Picket fences, usually shorter than 5 feet in height, are decorative and do serve to confine small animals but can be jumped by larger ones. A picket fence must be maintained. We also expect picket fences to be clean and painted. A cedar fence may weather and turn grey without shame, but a picket fence looks terrible without a coat of clean white paint.

Westerners are aware of the popular stick fences used there, made of thin branches or trunks of small desert trees. Such a fence always looks very good, no matter how long it has aged, and individual sticks are easily replaced when damaged. Such fencing is very expensive where thin desert trees do not grow, however.

Fence builders use a special quick-drying concrete in their postholes. This sets the post very quickly—in a matter of five minutes or less where the ground is dry—but allows no time for changing things around if an amateur fence builder changes his mind or makes a mistake. If you use the fast-drying concrete mix, place all the posts at once, and make sure the measurements are correct before pouring the concrete.

Don't forget about a gate or two. If you have a truck, camper, boat, or other large vehicle to store away in a fenced yard, make the gate wide enough to enter and exit safely and easily.

Some homeowners leave forested lots in their natural state or plan desertlike rock gardens that cover all the property. These are attractive ideas if you consider the fact that every special departure from the grass-lawn tradition will limit future sales of the home to that special buyer who agrees

This weathered rock, probably weighing a ton or more, would have been difficult to haul away, so the homeowner kept it as a lawn decoration.

with such a difference. A desert-landscaped lawn, covered with gravel and lava stones, saves yard maintenance, obviously, but it is not the place a family of playful children want to throw a ball in.

If you decide to use desert-type landscaping, however, be sure that the ground under the rock has been sterilized—that is, killed so that nothing will grow up through the stone and leave you with a lawn maintenance problem you did not expect.

If you opt for a natural forest setting, inspect every corner of the property for poison ivy or other bothersome weed, or animal pests such as gophers or termites, and carry out such inspections every few months so you will catch any new growths or invaders. Keep doors and windows closed and even your chimney flue shut, to keep squirrels and chipmunks out of the house.

A forested lot probably eliminates your chance for a garden. Whatever plants succeed in getting sunlight enough to grow will probably becomes lunches for forest animals long before harvest time. If you have a pet dog or cat, have plenty of flea and tick powder on hand.

But growers of lawn grass will have problems, too. The area where you build a home determines which type grass you can use. Red fescue, Kentucky bluegrass, and rye grass are popular in cooler climates, while Bermuda and St. Augustine are usable where it is warmer. Rye is a fast-growing, single-stalk grass, while Bermuda puts out trailers which, in turn, grow into more stalks. Many people are allergic to Bermuda which gives off a tremendous amount of pollen in late summer. All grasses need mowing from time to time, as well as weeding and fertilization.

There are low-growing dichondra grasses, similar to clover, and hybrids such as zoysia available for lawns, too. Some hybrid grasses do not grow from seed but must be individually plugged into the soil. Mixes of various types of grass are also available for quick coverage and long-term growth, but in time, the stronger type will probably overpower the weaker and leave you with one grass species and a lawn that turns brown all at once in fall, returns to green at the same time in spring, or is entirely wiped out by sudden frost or pest or blight. In using a mixture of lawn grasses, you should periodically rip out old growth and reseed.

Before deciding the grass choice for a new lawn, consider the amount of traffic you expect, your

A boulder like this one can be a decoration for your finished lawn if you want it, but make sure that the fellow with the tractor who comes to grade the property moves it for you. No one has enough friends to get this job done manually.

Lois Schild of Dayton used some imagination and sculpting skills to create this unique house number. She worries about theft, always a problem with such decoration or lawn furniture.

Plan your packing and moving carefully, especially with young children. Load toys last, so they'll be unpacked first.

local climate, and your preference for a tall, leafy grass or a short, tightly woven type. Look at what the neighbors have planted. If a neighbor has a hardy bluegrass on his lawn and you want a delicate rye, you may find bluegrass everywhere in a few years as the rye loses way. Your nurseryman will be happy to advise you.

Another consideration when planting a lawn and garden around your new home is how you will provide water. If you will be using hose and movable sprinklers, consider where the water faucets are before laying out a plot which requires daily watering, such as an herb garden. If the yard is big enough, it may be most convenient to spend the money and install underground sprinklers that can be turned on with a simple turn of the handle.

If you use sprinklers, consider whether or not the ground freezes in winter. If those sprinkler pipes, which are virtually impossible to empty completely, will freeze solid and burst in a long, hard winter, stick with the hose. In warmer climates, where watering is necessary to keep grass and shrubs alive against a broiling sun, the underground sprinkler will prevent the sun from broiling you, too.

A plumber can hook up an exterior pipe to feed your sprinklers, or you can tie into your exterior faucet line if it is conveniently located. Each sprinkler system has to be designed individually, but your local plumbing supply or hardware store offers galvanized or copper tubing fixtures to meet virtually every need.

Take measurements of each leg of tubing; count the number of elbows, tee fixtures, and plugs needed; count the number of sprinkler heads you need, and the direction in which each must squirt—half circle, quarter circle, or full circle—and take all your figures to your supplier.

Remember to get an on-off valve, too, for you do not want to turn on the whole sprinkler system every time you need a half bucket of water to wash the patio furniture.

Your supplier can probably thread each length of pipe for you, so the job that remains is merely to attach the pieces to one another. Use Teflon tape or joint compound on each joint for a secure, leakproof fit.

If you are using outdoor carpet, preglued tile, or artificial grass on any surface outdoors, follow the manufacturer's instructions on its installation. Use a little common sense in placing such expensive material outdoors, too, so that greasy lawnmower wheels, wet feet, hot coals, or greasy picnic treats do not mar it later.

If your lot needs landscaping and trees, consult a nurseryman in your area to find out what plants are most appropriate for the climate. If you are starting on a bare plot, you probably need a few fast-growing shade trees, such as willow, and some slower and more stately species, such as ash or oak. Fruit trees are fun to pick, but keep in mind that few trees ever look as good as the picture in the seed catalog, and the fruit must be protected from birds and animals. Berry trees and bushes can be pleasant but messy, too. Consider where that mature fruit will land when it falls.

Olive, mulberry, cherry and some nut varieties are very messy but attract many popular birds. Do you want birds, or will you be sorry you invited them? Don't wind up like the spinster sisters in Dayton who spend their summers on their front porch, clapping their hands at the flocks that nest in their colorful hackberry tree which overhangs the front walk and porch.

In all areas of the country, be wary of the door-to-door peddler who arrives with mature trees, to sell at ridiculously low prices. Such plants are often torn from the earth, without regard for their root system or transplant shock, and die soon after your check clears the bank. Deal with a reputable nursery.

In climates where citrus grows, land developers and subdividers sometimes offer mature orange, lemon, or grapefruit trees to those who haul them away. Such trees, if carefully extracted from the ground with minimal damage to the large roots, can be transplanted but require heavy pruning immediately after transplant, and lots of water. First new fruit may take several years. Find out how old the tree is before bothering with transplant, however; citrus trees are only efficient for about twenty-five years, and if they're already twenty years old, you will be wasting your time and energy.

Live Christmas trees are pleasant lawn inhabitants where the climate allows evergreens to thrive, but if your area is one of heavy freeze in December, you'd better plan far ahead and have the hole dug in November or even October before the ground gets hard.

When planting a healthy tree with its root ball wrapped as they come from the nursery, dig a hole twice as big as the root ball, and then refill it halfway, so that there is loose soil underneath the tree. Then fill the hole with water two or three times, letting all the water drain between fillings. If the roots are tightly tied with rope, wrapped in plastic, or in a pot of any kind, remove all plastic and loosen the rope slightly. Paper root wrappings or loose burlap deteriorate by themselves and need no unwrapping. Now insert the tree. Fill in the hole and water once again, thoroughly, and add fertilizer.

Where the soil is not particularly fertile, some nurserymen recommend throwing a few whole potatoes into the hole before placing the tree. Potatoes will grow there and, perhaps emerge as greens. Cut them off at the ground if they do, and let the vegetables rot and form organic matter for the young tree.

Avoid water-seeking trees such as tamarack whose roots may creep along underground far beyond the tree's "drip line" and invade your sewers, or your neighbor's. Also avoid any highly acid- or alkaline-loving tree if the soil in your area is not suitable, or you will spend many hours pumpting strange chemicals into the ground to keep some favorite tree alive. If the soil is not proper for a particular tree, no amount of artificial fertilization can ever make it thrive through maturity. At best, it will live a stunted, unsatisfactory, and short life. Magnolias grow in the warm and moist South, and apple trees grow in the frigid North. So be it.

What about shrubs? Small plants are excellent walk-edgers, foundation decorators or step-side sentinels. If you like flowers, ask your nurseryman about what is suitable for your

Moving-in day looks terrible. The walls are bare of decoration, and boxes, both full and empty, are strewn about. However, it's the start of your new life at this address, and so it is enjoyable.

climate. You can choose from annuals which grow from seed, flower, and die all in one year, or from perennials which last for many years. Annuals obviously need yearly replanting but include many beautiful blooms, such as pansies; perennials grow more slowly but include the earliest blooms, such as tulips or crocus.

Evergreens, such as juniper or arborvitae, grow in various high or low forms and are excellent accent plants. Roses are beautiful and smell sweet but require care to survive hard winters and must be nursed through various crises involving pests, fungus, or feeding. Once a rose blooms, the flower must be clipped, or that branch will not bloom again.

There are hundreds of choices to make when landscaping your home, and you must do it individually with consideration for the layout of your lot, the size and shape of the house, personal preferences of color and scent, where you want shade and sun, and where you want leaves, fruit, or flowers to fall when the time comes. It's really great fun to lay out flower beds, gardens, and hedges around a home you have seen grow from the ground up. When you reach this point in the progress of your house, you will be satisfied with your accomplishment and should be very happy with your new home. The next step is moving day.

Congratulations and good luck.

index

Page references in italics refer to illustrations.